Glendale Library, Arts & Culture Dept.

3 9 0 1 0 0 5 6 8 0 0 3 0 7

NO LONGER PROPERTY OF
GLENDALE LIBRARY,
ARTS & CULTURE DEPT.

D1006066

THE WISDOM OF
WOLVES

Also by the authors
The Hidden Life of Wolves
Living with Wolves
Wolves at Our Door
The Sawtooth Wolves

For children
A Friend for Lakota
Living With Wolves!

THE WISDOM OF
WOLVES

LESSONS *from the* SAWTOOTH PACK

JIM & JAMIE DUTCHER

WITH JAMES MANFULL

NATIONAL
GEOGRAPHIC

Washington, D.C.

599.773 DUT

Published by National Geographic Partners, LLC
1145 17th Street NW Washington, DC 20036

Copyright © 2018 Jim and Jamie Dutcher. All rights reserved. Reproduction of the whole or any part of the contents without written permission from the publisher is prohibited.

NATIONAL GEOGRAPHIC and Yellow Border Design are trademarks of the National Geographic Society, used under license.

ISBN: 978-1-4262-1886-6
ISBN: 978-1-4262-1994-8 (deluxe)

Library of Congress Cataloging-in-Publication Data

Names: Dutcher, Jim, 1943- author. | Dutcher, Jamie, author.
Title: The wisdom of wolves : lessons from the Sawtooth pack / Jim
    Dutcher and Jamie Dutcher.
Description: Washington, DC : National Geographic, [2018]
Identifiers: LCCN 2017033893 (print) | LCCN 2017054884 (ebook) |
    ISBN 9781426218873 | ISBN 9781426218866 (hardback)
Subjects: LCSH: Gray wolf--Behavior--Idaho--Sawtooth Wilderness.
    | Social behavior in animals--Idaho--Sawtooth Wilderness. |
    Human-animal relationships--Idaho--Sawtooth Wilderness. |
    Wildlife photography. | BISAC: NATURE / Animals / Wolves. |
    PHOTOGRAPHY / Subjects & Themes / Plants & Animals.
Classification: LCC QL737.C22 (ebook) | LCC QL737.C22 D882 2018
    (print) | DDC 599.773--dc23
LC record available at https://lccn.loc.gov_2017033893

Since 1888, the National Geographic Society has funded more than 12,000 research, exploration, and preservation projects around the world. National Geographic Partners distributes a portion of the funds it receives from your purchase to National Geographic Society to support programs including the conservation of animals and their habitats.

Get closer to National Geographic explorers and photographers, and connect with our global community. Join us today at nationalgeographic.com/join

For information about special discounts for bulk purchases, please contact National Geographic Books Special Sales: specialsales@natgeo.com

For rights or permissions inquiries, please contact National Geographic Books Subsidiary Rights: bookrights@natgeo.com

Interior design: Nicole Miller

Printed in the United States of America

18/QGF-PCML/1

*This book is dedicated to the many
hundreds of thousands of people, like you,
who join us in believing that
wolves have the right to live their lives
and to make their unique contribution
to the world we all share.*

CONTENTS

Lakota, the pack omega

FOREWORD

EACH AND EVERY WOLF has a story to share. Can we be trusted to listen?

In a time when humans are wantonly and brutally exploiting wolves and numerous other nonhuman animal beings, it is essential that we pay very close attention to what they are saying to us as they try to adapt to a world in which their interests are far too often and universally trumped "in the name of humans." It's pretty simple: We rule, other animals have to do what we want them to do or they suffer the consequences of our narrow and anthropocentric demands that seriously compromise their well-being and their very lives.

We are living in an epoch called the Anthropocene, or the "age of humanity." In our book *The Animals' Agenda: Freedom, Compassion, and Coexistence in the Human Age*, Jessica Pierce and I call the Anthropocene the "rage of inhumanity" because of the unprecedented and unrelenting ways in which humans all over the globe are decimating countless animals and their homes. We restrict their freedom, so that even so-called "wild animals" are not all that free.

The Wisdom of Wolves clearly shows why we must open our hearts to the plight of other animals. Jim and Jamie Dutcher have spent many years living with wolves, both watching and listening and also documenting their

Kamots, the alpha, bowing in play

unparalleled experiences in both film and prose. The unique opportunity of their being able to share the lives of the Sawtooth Pack has brought the story of wolves to people all over the world at a time when it's needed most. The intricate and intimate world of these iconic family-living North American animals, previously hidden from human eyes, has given us all the gift of new insights into who they are, how they live, and the depth and richness of their emotional lives.

The book you hold reveals a panoply of emotions, ranging from unbounded joy to excruciating sadness and grief due to the loss of their companions, including adult family members, their children, and their friends. It is vitally important for us to pay close attention to what wolves and other animals feel when they are able to live together peacefully and when they are subjected to hunting and losses of habitat due to human encroachment and climate change.

The world of wolves the Dutchers have carefully detailed creates a framework that is needed to enhance our attempts at the conservation of these magnificent animals. The more we know about their life histories, including patterns of births and deaths, how social bonds and packs are formed and maintained, and how human intrusions influence the lives of wolves, the better we will be able to protect them in a human-dominated world. Researchers and non-researchers alike will greatly benefit from carefully reading this book. If we lose wolves, we lose centuries of information we will need as we confront a changing world that is not especially animal friendly.

FOREWORD

We, as Earth's dominant species, need to learn as much as we possibly can, in as many ways possible, to comprehend our intricate interconnections with other animals and with nature as a whole, as unmatched and increasing threats engulf our world. The more we know about wolves, the more we know about ourselves, and the more we can pass on to our children. This is the definition of wisdom, as the title of this very important book aptly highlights.

Every wolf, every passing shadow in the forest, every individual watching us as we pass through her or his territory and home, indeed all wolves have a story we need to hear and to use on their behalf. Can we be trusted to listen? I sure hope so, because wolves and other animals need all the help they can get, and time is not on their side.

The Wisdom of Wolves will help us rewild our hearts, become reconnected to nature and to other animals, and work hard to build pathways of compassion and coexistence. If we don't foster and maintain deep and long-lasting respect, compassion, and coexistence, we will lose out as much as the other animals will. When other animals lose, we lose, because we suffer the indignities to which we subject individuals of other species. If we work hard for compassionate coexistence, it will be a win-win for all.

Marc Bekoff, Ph.D.
Boulder, Colorado
June 2017

INTRODUCTION

T HEY'VE DECIDED THAT THIS IS THE TIME. The sun has disappeared behind the mountain peak to the west, leaving just a wisp of orange cloud, stealing minutes from the short February day. In the east a deep blue creeps in, too early yet for stars. Night is falling still, clear and bitterly cold. They don't mind. They're creatures of the cold. They appear in near silence. One emerges from the shadows of the spruce forest, two from the tangle of bare willows. In the clearing, eight now circle one another. Old snow crunches under soft footfalls. There was no audible call to assemble, no visible trigger that sparked this gathering. They've got their reasons.

They exchange greetings, touch noses, press shoulder to shoulder, and then move apart. Their movements betray anticipation, something between agitation and joy.

In the apparent chaos of the assembly there is a subtle order. One stands taller and more still than the rest, like the hub in the center of a wheel. He's not the focus of the action, but perhaps its core—fussing less, confident, alert, solid. Each individual makes a point of greeting him first, though their level of deference varies from individual to individual: One approaches in a relaxed and respectful greeting, another approaches in a bow so deep he's almost crawling—the picture of genuflection. Here in the clearing,

Matsi and Kamots, beta and alpha wolves

personalities are laid bare. These are individuals, each with his or her way of being and interacting with the others. There are displays of boldness, timidity, aggression, playfulness, solemnity, and tenderness.

All the while they've been speaking quietly, sharing something between them, in voices resonant with excitement. Suddenly the leader's voice soars above the chatter. His call rises quickly then slowly falls in tone, tapering off where it began. For the others it's the cue to begin. Each singer adds his own voice to the chorus but sings a song all his own. The sound is both dissonant and harmonious. It fills the valley below, echoes off the mountain walls, and floats back in a doubled loop. With faces turned to the sky, they sing. They are a tribe, a family, a fierce confederacy. They are also an assembly of individual personalities, private desires and goals and inner lives largely unknown. They are a wolf pack howling.

In Alaska and Alberta, in Wyoming and Wisconsin, in Germany, Spain, Russia, and the Middle East—anywhere on Earth where *Canis lupus* lives, this scene is surely repeated. Sometimes we hear it; rarely do we observe it. But this particular time, it's a bit different. The setting is certainly prime wolf habitat, and the wolves present have gathered for their spirited pack rally for reasons known only to them. However, they are not completely alone. From the darkness of the bordering forest a pair of human observers crouch at a discreet distance, watching, recording, marveling at the spectacle they've been privileged to witness.

Back then we were known as Dutcher Film Productions—Jim Dutcher on camera and Jamie Dutcher recording sound—documenting events that hadn't occurred here for the past 50 years. We called them the Sawtooth Pack, after the majestic Sawtooth Mountains that rose behind their riparian home on the edge of the Idaho wilderness. For six years they allowed us to share their lives, and we came to know them both as individuals and as integral parts of their pack. We were passive observers in their lives and honored to have had their trust.

The calm center of the Sawtooth Pack was the benevolent alpha male we called Kamots, a Blackfoot word meaning "freedom." He was a strong and confident wolf but was not above playing with his packmates and indulging in boundless curiosity. His mate was a high-strung dark female we named Chemukh. On the opposite end of the pack hierarchy was Kamots's littermate, a big gray male we called Lakota. Despite his size, Lakota was timid and shy. For many years he was subjected into the unenviable position of omega, the lowest member of the pack, often enduring aggression from the others. Five other wolves—four males and one female—made up the pack's middle ranks, and in 1996 Chemukh gave birth to three pups—a total of 11 wolves.

The Sawtooth wolves weren't exactly wild—but then again, they couldn't have been. At the time there were virtually no wild wolves in Idaho or throughout the American West for that matter. We knew, however, that plans were under way to return gray wolves to a portion of their former range, and it seemed the right time to produce a film about these animals. Our goal wasn't to document every aspect of wolf behavior; it was to show the intimate lives and relationships within a pack. In two films—*Wolf: Return of a Legend* and *Wolves at Our Door*—the Sawtooth Pack helped reintroduce Americans to wolves, just as their wild cousins were being reintroduced onto American soil. In 1995 and 1996 the United States government captured 66 wolves in Canada and released them into Yellowstone National Park and Idaho's Frank Church–River of No Return Wilderness.

As the wolves' return into the Rocky Mountain states progressed on schedule, knowledge and understanding about wolves lagged far behind. There had been a few scientific studies, most notably in Alaska's Denali National Park and Preserve and on Ellesmere Island off the coast of Greenland, but public perception was still based more on centuries-old mythology than on any real observation. To some, wolves were demons to be feared and hated, vermin whose eradication has been justified (even divinely sanctioned) and whose impending return was tantamount to an act of domestic terrorism. To others, they were the spiritual embodiment of America's lost connection with nature—a pure and infallible creature more symbol than substance.

Initially the mythology surrounding wolves is what made them such an attractive subject for our filmmaking. In the early 1990s wolves were a hot topic, their mythic significance exaggerated by the plans to bring wolves back into the American wilderness. But it didn't take long for us to become captivated by these animals for altogether different reasons. The more we observed, filmed, and recorded their sounds, the less we saw the mythical beasts of our dreams and nightmares. What we found instead was something a lot more familiar.

Wolves can be described in many ways, but above all they are social. They need each other. As hunters, as parents, as keepers of a home territory, wolves succeed as part of a group. They've evolved to live and function within a society: They communicate, cooperate, teach their young, and share the duties of day-to-day life. How could they achieve such social sophistication without developing and maintaining strong interpersonal connections? In six years observing and filming the Sawtooth Pack, we have come to see wolves as deeply emotional individuals who care about what happens to themselves, family members, and friends. Like many other animals, they are emotionally intelligent beings. A wolf knows who he is, and he sees his packmates as individuals. He has a concept of how his actions are perceived by others. He is capable of empathy, compassion, apology, and encouragement.

As humans, we're empathic creatures too. So many times, watching the Sawtooth Pack, we were struck by the realization that we weren't just sharing the wolves' experiences, we were sharing their feelings. Our joy at the first snow of the season was mirrored in the way they raced about, playing tag and rolling in the fresh powder. When Chemukh gave birth to pups in a den she had dug under a fallen tree, we were over the moon. We couldn't wait to see them and to make sure they were healthy. As we watched the pack gather around the den entrance, pacing and whining with excitement, we could see they had exactly the same feelings we did. When a mountain lion killed a member of the pack, the change in their behavior was undeniable, and we felt their grief.

Our observations have only been bolstered by the conversations we've had with wolf biologists from Yellowstone to Alaska. As we were beginning our work with the Sawtooth Pack in the 1990s, we had the opportunity to visit and film wolf biologist Dr. Gordon Haber in Denali. At that time he was conducting the longest continuous study of wolves in the world, begun in 1939 by Adolph Murie. When Gordon Haber died in a plane crash in 2009, we lost a meticulous scientist and an outspoken wolf advocate. Few human beings have logged more hours observing wolves than Gordon. He was a rigorous scientist, as evidenced by notebooks filled with minute-by-minute observations. Yet he was not afraid to step beyond pure science and acknowledge that wolves resonate in the human psyche in a way few other animals do. This deep connection with wolves represented the underpinning of all the important scientific work Gordon Haber did. He expressed that awareness himself in his book, *Among Wolves:*

> In many ways the social organization of wolves is surprisingly similar to what anthropologists have pieced together as the social organization of early

man. The well-defined dominance order and the disciplined manner in which duties are assigned and carried out, the presence of different generations of the same family living together, the prolonged dependency of the young, the group effort in raising and training them, the cooperative effort of many individuals in hunting large prey—in these and other respects, wolves, like our own human ancestors, have developed a highly effective means of coping with a wide variety of ecological conditions. Perhaps by more intensively studying—not persecuting—a species so similar in behavior to that of our ancestors, we can learn much more about ourselves.

It is now understood that all domestic dogs are the descendants of wolves, and there is good evidence that domestication likely began more than 15,000 years ago, possibly over 30,000. Yet whenever domestication occurred, our special relationship with wolves certainly began much earlier. Many hypotheses suggest that ancient humans and wolves lived in close proximity and may even have hunted together. Humans and wolves were, and still are, two of the most geographically widespread mammals on Earth. As our ancestors were spreading across the globe, wolves were there. As two of the few animals to live in extended family groups where cooperation is essential, it's not surprising that we became kindred spirits. In the lives of wolves we see reflections of things we have come to value in ourselves. They care for their pups with a familiar devotion and share our reflexive instinct to care for youngsters, related or not. They hold a place in society for their elders. They push boundaries and explore, then return to visit their families. They care what happens to one another, they miss each other when they're separated, and they grieve when one among them dies. They do all of this with a clarity of purpose that we often struggle to achieve.

One can't observe a wolf pack without seeing a reflection of ourselves, although the reflection is not always unconditionally positive. The life of a wolf can be fiercely complicated. An individual wolf has bonds to its pack but may also harbor a strong personal drive for breeding and territory. The tug-of-war between society and individual ambition adds a layer of complexity that occasionally plays out in Shakespearean-style narrative. It's one of the things that makes wolves endlessly fascinating and personally compelling.

The wolves of the Sawtooth Pack gave us not only the rare opportunity to share their private lives but also a greater understanding of our own human nature. Their lives and behaviors mirrored our own contradictions and complexities: social hierarchy tempered by compassion, contention mixed with cooperation, the admirable side by side with the abhorrent. Among all their qualities there were many we admired, but one stands out more than any other, especially considering all that has befallen them at the hands of man. Wolves can forgive.

INTRODUCTION

Rick McIntyre, park ranger, biological technician, author, and friend of ours who has come to know every wolf in Yellowstone since reintroduction, recently quipped in an interview, quoted in Carl Safina's book *Beyond Words:* "Certain wolves I've known—they were better at being a wolf than I've been at being a person." We can't count the number of times we experienced that sentiment while watching the Sawtooth Pack.

It is with this spirit that we write this book. Our intention is not to anthropomorphize wolves or to imbue them with human morals; it is to celebrate their very wolflike qualities through the lens of our own humanity. As it happens, many of the qualities that make a wolf successful at being a wolf also represent the best in human nature. In this book we have gathered favorite memories from our years with the Sawtooth Pack as well as stories of other wolves we have come to know. They are benevolent leaders and faithful lieutenants, fierce mothers, nurturing fathers, and devoted brothers; they are hunters, adventurers, comedians, and caregivers. Wherever there are wolves, there are tales as inspiring as any human narrative. We owe it to wolves to let them live their stories without persecution and without judgment. We owe it to ourselves to watch and marvel and reflect. Our lives are richer when we listen to what the wolves have to teach us.

Note from the editor: Chapters in this book are written in either Jim or Jamie's voice, allowing each of them to share personal observations and experiences. You will see the signature of one or the other at the start of each chapter.

CHAPTER ONE

EARNING TRUST

JIM

THE SOUND SEEMS TO COME from everywhere at once. Outside there is nothing but the freezing black of a moonless January night. Jamie and I peer out into the darkness, but we see no movement, no source. From the deck of our camp, we can barely make out the surrounding mountains as snow clouds spill in from the west and we stand listening.

The howl grows more intense. For a moment I think I can pick out individual voices coming from a definite direction, but then they dissolve and meld into a dissonant chorus. At times it sounds like there could be 20 of them. We know there are only eight. But what a sound these eight can make! It's eerie and mournful, exuberant and mysterious—all the contradictions of the American wilderness wrapped up in one sound. Looking at Jamie, I know it triggers the same conflicting emotions in her as it does in me—a strange mixture of loss and hope, a feeling of connection, a great sense of privilege, and maybe just a tinge of primordial awe.

Kamots howling at wolf camp, Lakota and Wyakin in view

After a few minutes the song dissolves. Voices drop out one by one. Somewhere in the dark there's a single final *oop* from a yearling wolf, still perfecting his howl, and then everything is silent. The Sawtooth Pack has just treated us to another pack rally, culminating in a riotous howl. We will never know precisely why they do it—only they know their reasons—but if we could stay with them for 20 years, it would never get old. Jamie and I exchange smiles and I squeeze her hand, acknowledging the precious experience we have just shared. Then we go back to the business of making dinner in silence.

WE CALLED IT "WOLF CAMP," a deceptively simple name for the little patch of tents, platforms, and fencing we had constructed. Wolf camp was more than a place; it was the embodiment of all the inspiration, trial and error, and years of hard work that had gone into this project. It was the foundation of the entirely new way that we were telling the world about wolves. The project began as an idea: Find a way to share the hidden lives of animals who were so scarce and so elusive that they had become virtually unknown. The project was to span a few years and culminate in a documentary film. It evolved into something far bigger than we ever could have imagined.

What became our life's passion started in 1990 as an idea for a one-hour TV documentary. My film *Cougar: Ghost of the Rockies* had just aired on ABC Television's *World of Discovery*. Before that I had made *Rocky Mountain Beaver Pond* for National Geographic. It felt like I was on a roll, and I was eager to find my next subject.

I wasn't sure what it would be, but I already knew how I wanted to approach it. Many years earlier I had read the work of Austrian zoologist Konrad Lorenz. One thing he wrote had stayed with me all that time, inspired me in my filmmaking, and has since informed everything I do. He wrote, "The quickest way to learn the language of a species is to do so as a social partner." With that in mind, I made it my mission to give the audience of my wildlife films a personal experience with my animal subjects. I specifically chose animals who were very difficult to see, let alone film. Beavers aren't hard to locate, but they have the frustrating habit of hiding out in their lodge most of the day. When they do venture out on land, the slightest disturbance sends them dashing for their pond. Cougars are the opposite. They're unafraid, but elusive. Catch a glimpse of one and you may never see it again. So I had to come up with original ways to get close to these creatures, either by filming animals habituated to my presence, in enclosures, or with the beavers, in our authentically constructed cutaway beaver lodge, to bring viewers in close for an intimate look that they wouldn't get any other way. I knew that after *Cougar* and *Beaver*, my next film would have to do the same.

As it turned out, I didn't find my next subject; it found me. On a hike in Wyoming's Absaroka mountains I crested a ridge and glimpsed what at that time was one of the rarest sights in the lower 48 states: a wolf. It was crossing a high alpine meadow more than a hundred yards away, but it saw me the moment I saw it. By the time I brought my binoculars to my eyes, the wolf had darted over the next ridge.

Here was an even greater challenge than the previous animals I had sought to film. Wolves weren't just elusive; they were very nearly gone from this part of the world. The wolf I had seen was probably a young disperser, a wanderer who had left its pack and gone looking for a mate and new territory. It likely found no others and turned back for home—possibly in northern Montana or perhaps across the Canadian border. I found myself wondering about its inner life and its quest for companionship. I wanted to know this animal.

Immediately I began thinking about how to approach making a film about wolves. I recognized that they are the most misunderstood large mammals in North America, which made the challenge even greater. If I could bring my audience in close, give them a personal experience, and show them the animals' inner life, people would understand and care.

To achieve that goal, I had to create a setup similar to what I had done filming beavers and cougars: a semiwild situation where the wolves would be accustomed to my presence and allow me to film them without changing their behavior. I would be filming in an enclosure, but I wanted to be as transparent as possible and make the project part of the story.

A primary reason for this method was purely practical: There was simply no way to film truly wild wolves at a close range to achieve the depth I was looking for. I had seen what others had achieved after spending months in the field in Alaska and Canada, and the results were admirable but not at all intimate. Jamie and I had filmed wolves in Alaska, and I began to refer to the faraway creatures as *Canis lupus minisculis*. They were always just tiny flecks in the lens. With the help of a very long telephoto lens, I was able to capture them moving about, occasionally playing or hunting (if I happened to be at the right place at the right time), but I wasn't terribly interested in such a limited story. I wanted to know how they interacted, communicated, showed affection, and resolved conflicts. To do that, I had to see into their eyes. I had to be close enough to detect the nuances in their behavior and hear the sounds they made to one another. If I tried to get that close to wild wolves, I wouldn't have seen anything but their tails disappearing into the trees.

I had deeper reasons for setting up the wolf project the way I did. By 1990 most conservationists knew of the federal plan to restore wolves to some of their former range. Five years before reintroduction, anger and resentment were already brewing. The more extreme wings of cattlemen's associations and hunting groups had already pledged to "shoot, shovel, and shut up"—in other words, shoot and quietly bury any wolf they could. In that climate, it would have been irresponsible for me to habituate wild wolves and get them comfortable being around a man with a big camera. The next time someone pointed something at those wolves, it probably wouldn't be a camera.

Thus I set out to film North America's most controversial animal at the very moment when that controversy was reaching a boiling point and in the region of the country where the anti-wolf movement raged with the greatest fury. Suffice it to say, I took precautions. If I had started the process a year later, the anti-wolf opposition would have had time to organize and would have shut down the project. As it happened, I didn't fully grasp the scope of all the hatred, but my timing was perfect. My plan for creating an environment that would allow me to film wolves was emerging.

THE WOLF PROJECT WAS AMBITIOUS: To create a safe, natural habitat for wolves that allowed me uninterrupted filming, I needed to find the perfect spot. I required a location that was accessible to an equipment truck to ferry in fence materials but remote enough that strangers could not easily find it. I needed access by all-terrain vehicle in the wet months of spring and by snowmobile during the long months of winter. Not only did we need to bring in all our food and the appropriate film gear and equipment, but also food for the wolves—usually road-killed elk or deer that we collected for them. I needed a place that was picturesque and had a variety of vegetation and topography in which to film. For the wolves, it needed to have a good balance of sunlight and shade and a consistent water source that wouldn't dry up in the summer.

So with my small crew of two, I started searching for the perfect location. I knew we were going to find what we needed only on Forest Service land—private land bears too many signs of human development, so to find land still wild and natural, it was going to be land under federal protection. We made 17 trips into the backcountry on foot, by four-wheel drive, or by snowmobile. We looked into all the surrounding mountain ranges: the Pioneers, the Boulder/White Clouds, and the Sawtooths. One morning in early October 1990, after following an overgrown jeep trail behind a ranch and hiking a short distance into the foothills of the Sawtooths, we found ourselves standing in the middle of a wide, grassy meadow. To one side ran a mountain stream, to another stood a grove of aspen. In front of us was a deep forest of lodgepole pine and spruce, out of which soared the gray walls of the Sawtooth Mountains. It was absolutely perfect.

Next came the lengthy process of obtaining permits to operate on Forest Service land, a process I was familiar with from my previous films. To keep the project as secret as possible, I hired an out-of-state fencing company to construct a 25-acre chain-link fence, 10 feet high with a 6-foot apron facing in and an overhang in all the corners so that even the most determined wolf couldn't dig under or climb over it. After a year of preparation, the only thing missing was a pack of wolves.

I suppose I was a little naive in those early months. Under pressure of a filming schedule I thought I could jump-start a pack by borrowing adult wolves from two rescue centers: a male we named Akai and a female named Makuyi, the first official Sawtooth wolves. I soon came to understand that a wolf pack is not something a human being can simply assemble at will.

Makuyi was a beautiful, sweet wolf, and I was taken with her the moment I first set eyes on her. Unfortunately, as I was soon to discover, she suffered from cataracts, and her poor vision made her spook easily. The following year, when a litter of pups joined the adults, Makuyi became increasingly fearful and spent most of her time as far away from the others as she could get. We brought in a skilled veterinary ophthalmologist and performed a successful corneal transplant—possibly the first such operation conducted on a wolf. I believed that restoring her vision would enable her to integrate with the rest of the pack, but wolves don't work that way. Her status as a semi-outsider had already been established. And so rather than subject her to a life alone, I returned her to her original caretakers, where at least she would be in familiar company.

The male was a big photogenic wolf named Akai, who certainly looked the part of alpha. Like Makuyi, Akai had grown up at a wolf center and was used to people. I came to learn, though, that a wolf's being used to people is not the same thing as a wolf's trust. My project required me to get close to these wolves, to film them feeding and during displays of dominance when they were sorting out their pack hierarchy. But Akai didn't know my crew or me—or even Makuyi. As a result, he was unpredictable. He was usually fine around people, but occasionally something would frighten or anger him. He never injured anyone, but there were times he seemed as if he could have. I simply did not feel that Akai could be trusted, and probably Akai felt the same about me. It was a frustrating start, but today I look back at the learning curve of this period and realize that some of these early mistakes were what ultimately made the wolf project so successful. For wolves to live together in a pack, they must trust each other without question. For me to be able to live with these wolves, and film their lives intimately, they needed to trust me.

My knowledge of raising wolves evolved out of research far and wide and, most important, from an association with the wolf expert Erik Zimen, who had raised a captive pack in Bavaria. Erik himself was a graduate student of Konrad Lorenz, who had been such an inspiration to me. I met Erik at a wildlife film symposium in Bristol, England, and he stressed to me that trust was the most important ingredient in the formation of a safe and healthy pack—trust that could be achieved only by living with the wolves as social partners. This was a new idea for me, and a challenging one. I had hoped that, as with the cougars, I could simply let these animals live their life without interference and they would reveal all their secrets to me. But cougars and wolves are very different creatures. Every expert I spoke to said or implied the same thing: These wolves must trust me absolutely or my work would be impossible. I was going to have to be more involved in their lives than I had anticipated. This meant that the wolves would need to get to know me from the moment they opened their eyes, so it was essential that we start with pups.

As close as we got to the wolves, we never treated them like pets. The pups kept their wild ways, wrestling for supremacy and sorting out squabbles on their own. More than anything, wolves, like our beloved dogs, are social creatures. As pups they need companionship, a sense of belonging, and the bonds of a pack.

My original intent had been that this litter, and another scheduled to come to us a year later, would bond with the adults, Makuyi and Akai, but I soon realized that the adult wolves would never be the social partners the young wolves—and I—needed. The project rested solely with the pups that trusted us completely. I decided to return the adults to sanctuaries and relaunch the project with just the young wolves. Those two litters of six pups, born a year apart, formed the core of the Sawtooth Pack that, since then, legions of fans have come to recognize and follow.

THE FIRST LITTER, BORN IN 1991, produced the pack's two most famous faces: the brothers Kamots and Lakota. Their markings were classic wolf: mottled gray fur with a charcoal gray mask, a light muzzle, and amber eyes. As pups, and as adults, the two were difficult to tell apart from looks alone, but their behavior left no doubt as to who was who. Because wolves had long held a place in the culture of many Native American tribes, we had decided to use words from various indigenous languages to name these wolves. We called one of the new pups Kamots, the Blackfoot word for "freedom." He was a confident and curious young wolf. He held his head and tail high, and moved with a sure-footed lope toward any sound or perceived danger. By their second year, Kamots had solidified his position as the alpha—the top of the pack hierarchy.

His brother was just the opposite. We called him Lakota, which means "friend" in the Lakota Sioux language. He was definitely friendly but also timid and insecure, and his lack of confidence meant that the rest of the wolves sometimes picked on him. When there was any current of aggression within the pack, Lakota often bore the brunt of it. He was also playful, though, and was usually the one to instigate a game of tag. It seemed to be in his best interest to keep the pack's mood light, and he did this beautifully.

In this first litter there was also a shy, mysterious female with a coat of black and brown, and piercing yellow eyes. We gave her the name Motaki, the Blackfoot word for "shadow." Motaki was always the first to back down in any confrontation with her siblings—even more than Lakota—but she was also playful. For a time the wolves treated her as the pack omega, at the bottom of the hierarchy. She was timid, but she was also the one who could work the pack into a playful mood and get a game of tag going. Like Lakota, she could keep the pack's mood light. Still, there were times when the other wolves would get a bit too aggressive, and when that happened, Motaki would go off by herself.

I often wondered if she would ever be able to rise out of her omega stature, but I would never find out. One June evening when Motaki was just two years old, a mountain lion scaled the tall fence at wolf camp. The wolves' territory was large enough that the mountain lion could attack silently without the other wolves or humans being aware. It was the only time that the fence was ever breached, but gentle Motaki's death shook us to the core—as it did the others wolves. That incident convinced me that wolves feel the pain of loss as sharply as we do. They grieve and they mourn and they miss each other. Of that I have no doubt.

The second litter of pups joined the pack just after the loss of Motaki. In looks and in personality they were as different as three wolves could be. We called one of them Matsi, the Blackfoot word for "sweet and brave," and he fit his name perfectly. Of all the wolves he was the lightest in color. His fur was almost beige, and his face bore only a delicate mask. He had a sunny disposition to match his looks. He was seldom aggressive, and he didn't stoop to fights over who ate first or who was slightly higher in rank. He just kindly and gently assumed the role of beta wolf,

the second-highest-ranking member of the pack, dominant to all but the alpha. When we added new pups to the
pack a few years later, Matsi took it upon himself to look
after them. We also began to notice that Matsi subtly protected Lakota, who had been pushed into the omega position after Motaki's death. Matsi was a peacemaker, a
puppy-sitter, and a friend to all.

The second of the trio was a dark wolf we called Motomo.
As a young wolf he was almost completely black, and as he
aged, his fur took on a dignified grizzle. His name was also
a Blackfoot word, meaning "he who goes first." Motomo
didn't actually go first—that was Kamots. But he didn't go
last either—that was Lakota. He just sort of did everything
on his own terms. Strangely, Motomo seemed to take a
keener interest in our human activities than some of the
other wolves did. He was not interested in direct contact
with us; he was just very observant. When the rest of the
pack was out of sight somewhere, we'd notice Motomo sitting alone, watching us chop firewood or shovel snow with
his keen yellow eyes, as though he were trying to figure us
out. He was cool, smart, and enigmatic. Jamie often called
him "a man of few words."

The last of the new litter was a flamboyant wolf we called
Amani. His name was also a word from the Blackfoot
language, meaning "speaks the truth," but it wasn't a very
fitting name. He was a showman full of bluster and bra-
vado. Unfortunately, when he wanted to show off, he often
did so by picking on Lakota, straddling and nipping at the

beleaguered omega. For lack of a better word, he was the bully of the pack. On the other hand, when pups were around, Amani could be an absolute clown, indulging their every whim. A bully one moment, playful and endearing the next—Amani definitely embodied the complex and often contradictory inner life of a wolf.

WITH THE SAWTOOTH PACK STABLE AND HAPPY, I found myself making the biggest and happiest change in my own life. On a cold day in late December 1993, I picked Jamie up at the airport and whisked her off into the Sawtooth Mountains to sleep in a tent next to a pack of wolves, and to be by my side for the rest of my life. It had begun with a simple smile seven years earlier, boarding a flight to Washington, D.C., from Heathrow Airport. The smile in question belonged to a striking young woman with jet-black hair, dark eyes, and a backpack slung over one shoulder. I noticed she wore a beaded necklace distinctive of eastern Africa, so I asked, "Have you been in Africa?"

"Yeah, what gave it away?" she asked with good-natured sarcasm.

Then she flashed that smile again. Thus began the richest partnership of my life. Although we were not sitting together, we talked a bit during the long flight. I told her I had been visiting my sister, who operated a tented camp in a remote area of the Maasai Mara in Kenya. She said she had been in Zimbabwe, on safari with a girlfriend, touring the national parks and photographing wildlife. I described the thrilling time I had spent photographing cheetahs as they hunted gazelles on the savanna. It was clear she was as passionate about animals as I was—in fact, she had just taken a job in the veterinary hospital of Washington's National Zoo. It took me until we were at the baggage claim area to summon my courage and say I would be back in D.C. two weeks later—could I give her a call? To my surprise, she said yes.

That spring and summer of 1987, I flew from Idaho to Washington every month to finish the beaver film. Each time I visited, Jamie and I met for dinner. Our conversations were lively and engaging as we tossed around ideas for future films, but I always felt that something was holding me back. I was still suffering from the sting of a divorce, and I thought I was not interested in marriage again. There was the issue of my two children and the age difference between us, and there was also Jamie's established East Coast life. And so it did not come as a surprise to me when Jamie called at the end of summer to say that she had become engaged. I listened to her words from 2,000 miles away with the detachment of a condemned man. My entire being told me that I should say something—anything—to make her change her mind, but it seemed too far gone, with too many obstacles and so much distance. Years passed, and everything changed, in her life and in mine. Seven years after that first chance meeting, our shared Sawtooth Pack adventures began.

From the moment Jamie arrived at wolf camp, her excitement was infectious. She kicked off a brand-new phase of work, a brand-new camp, and additions to the pack. This city girl from Washington, D.C., jumped in with both feet! Her arrival inspired me to reinvent wolf camp and begin an entire new phase of filming. My goal had always been to capture the wolves' natural behavior on film by getting them as used to my presence as possible. Until that point, my tented camp was located next to but outside the pack's domain, which meant I had to enter a double gate to film them. Wolves are incredibly curious and inquisitive. Although I tried to be inconspicuous, whenever I came through the gate, laden with gear, they would stop whatever they were doing to come investigate. Their curiosity was endearing, but it interrupted their natural wolf behavior.

The solution was to end all the comings and goings and to merge our world with theirs. Essentially we created a small enclosure for humans within a vast territory for wolves. To keep the wolves from stealing our gear, we surrounded our camp with chain-link fencing. At the camp's center stood an eight-foot-high platform upon which we erected a yurt—a large, round tent originally used by nomadic horsemen in Central Asia. With our human activities in full view of the wolves, we quickly became uninteresting. The platform itself provided a great vantage point for observing and filming without their even noticing what we were up to, and when we slipped quietly through a gate and into the pack's territory, they often didn't even bother to look up—which is exactly what we wanted. For me, this merging of our two worlds was the fulfillment of the Konrad Lorenz quote that had inspired me many years earlier. We were no longer observers of wolves. We had become their social partners, and they opened up their lives to us as never before. From that point forward, every second of film, every still photograph, every observation we made was richer and more intimate.

As I enjoyed unprecedented filming opportunities, Jamie developed a passion for recording the language of wolves. We knew that for people to care about wolves, they had to see them up close. We came to realize that it was equally important to *hear* them. Jamie's pursuit grew into a passion. She delighted in the way the changing atmosphere of the mountains had such a profound effect on the sound of wolves howling, and she set about to capture all the different moods: flat howls on a foggy day, howls that echoed off the mountains on a clear winter night, the lovely dissonance of a group howl, and the mournfulness of a wolf howling alone. Of course the wolves presented her with a repertoire of other sounds that one would never hear without the benefit of proximity and a boom microphone. As the wolves spoke to one another, Jamie amassed a huge library of growls, whines, snarls, yips, and sounds she categorized as "Chewbacca noises"—their own verbalizations, uncannily similar to those of the Star Wars character. They always put a smile on our faces. Jamie's meticulous recording of their language deepened the intimacy that we were able to achieve. I can't imagine how we'd talk about wolves without that part of the story.

With the new momentum that followed Jamie's arrival, we decided it was time to introduce some new characters into our group of five males. Most important, the pack needed females to be complete. We brought three new wolf pups into the pack, two female and one male. By this point we had reached an agreement with the Nez Perce Tribe of northern Idaho to provide a permanent home for the pack once our Forest Service permit had expired. The Nez Perce were also slated to oversee the reintroduction of wolves to Idaho, so when it came to naming these three new wolves, we decided to use words from the Nez Perce language.

Two of the pups were siblings, a male and a female. To the girl we gave the name Wyakin. In Nez Perce lore, a *wyakin* is a spirit guide that appears to children, teaching them life lessons and special songs. Our Wyakin was a feisty spirit, lively and full of mischief. She loved nothing more than rough-and-tumble play with her brother, Wahots. His name meant "likes to howl," which seemed fitting for a wolf. He was a little more reserved than his sister, and later in life he relieved Lakota as pack omega. Even as they grew into adulthood, Wahots and Wyakin remained inseparable. The two were more than siblings—they were fast friends.

The third pup was a black female. As we struggled to think of a name for her, a friend started calling her Black Lassie. Of course that wouldn't do at all. In Nez Perce the word for black is *chemukh-chemukh,* and the word for girl (or lass) is *ayet.* Chemukh-Chemukh Ayet was not a name that rolled off the tongue, so we simply called her Chemukh. She was born to different parents than Wahots and Wyakin, and she had a very different temperament. She was timid, like Lakota, but where he merely lacked confidence, she was outwardly nervous and skittish. Early on, Wahots and Wyakin keyed in on her insecurity and often made a game of picking on her. Thus it was to our surprise—and probably to hers—when two years later she came into estrus and Kamots mated with her, making her the alpha female of the pack.

Chemukh never assumed the alpha role gracefully. With newfound authority she got her revenge on Wyakin, paying back the other female's prior harassment and more. The hierarchy of a wolf pack is generally divided along sexual lines—an alpha female is dominant over other females and an alpha male over other males—but Chemukh crossed the line and relieved her insecurity by picking on Lakota, a male three years her senior.

Despite her shortcomings, Chemukh entered motherhood with the assurance of a proper alpha female. In a den she dug herself, in a location she had selected for reasons known only to her, Chemukh gave birth to three pups. By all accounts they were the first wolves born in the Sawtooth Mountains for more than 50 years. All three were dark, like their mother: a male, Piyip, and two females, Ayet and Motaki II. For us, and for the viewers of our film *Wolves at Our Door*, the birth of these three pups was timely. Just one year earlier, in 1995, the U.S. Fish and Wildlife Service had released the first wolves back into Idaho. These reintroduced wolves were probably digging dens and having pups a hundred miles north of us that very same spring.

As a new act was beginning for wolves in America, the curtain was coming down on our wolf project in the Saw-tooths. Our Forest Service land use permit was expiring, yet we planned to provide for the Sawtooth Pack for the rest of their lives. It was against the law to set them free, and, more important, they had lost what they needed most to survive in the wild: a fear of humans. On an August evening in 1995 we loaded eight carefully sedated adults and three wide-awake, curious pups into transport crates. In the cool of darkness we made the seven-hour journey north and carried them to their new home.

WOLVES SELDOM LIVE LONGER THAN 10 YEARS. It is now more than 20 years since we released them into their new home. The visits we made from time to time were full of joy and tears in equal measure. The wolves always remembered us and rushed to greet us as old friends, whining, sniffing, and showering us with licks, even awkwardly trying to sit in our laps, perhaps remembering their days as pups a long time ago. As years went by we would receive heartbreaking news that one had passed, then another: feisty Wyakin, fearless Kamots, gentle Matsi. Piyip, one of Kamots and Chemukh's three pups, was the last to die, in 2013. In one day 98,000 people shared his passing with us over the Internet. Emotional messages poured in. The Sawtooth Pack had touched the hearts of people around the world.

We had given them names, but those were only for human ears. We never attempted to teach the wolves their names or call them to us. We only had contact with them if they initiated it. Often they did choose to approach us, and we would sit quietly, welcoming their company. They were free to ignore us or engage with us on their terms, just as they were free to sort out their own issues, establish their own hierarchy, and simply be themselves. This made the moments of connection that we did have with them all the more precious.

We will never know for sure how they looked upon us. Were we members of their pack or familiar friends? We asked them to be ambassadors for their kind, to help people understand their ways. We asked them to speak for past wolves who had been feared and persecuted, and for new wolves taking their first steps into an uncertain future. The Sawtooth Pack rose to this challenge with more grace than we could have hoped for. We gave them our pledge that we would care for them, respect their space, and honor their message. In return they gave us their trust. We will never forget that gift.

CHAPTER TWO

FAMILY FIRST

JAMIE

THE RHYTHMIC BEAT OF A RAVEN'S WINGS stirred my senses. The sound faded into the whisper of wind in the treetops and the gentle rustling of canvas. Slowly I began to remember where I was. I opened my eyes to the dull morning light and the familiar interior of our tent: the oil lamp on the small table beside our cot, the woodstove dark and cold. I could see my breath, but the cold wasn't as biting as it had been in weeks past, and the air had the slight smell of damp earth. It was late April in Idaho—a restless in-between time when winter was over but spring hadn't shown up yet.

Next to me, Jim was beginning to stir. I turned my attention to the world outside and listened for the sounds of morning at wolf camp—paws crunching through snow and cheerful voices speaking an unknown language. Since we'd moved our tented camp into the wolves' territory, we'd grown accustomed to hearing them as they shook off the chilly night and said good morning to each

Sawtooth pups Piyip and Ayet

other just a few yards from where we slept. Strangely, this morning I heard no such sound. I slipped out of bed and opened the tent flap, looking toward the edge of the forest where the pack usually spent the night. There were depressions in the snow where they had repeatedly bedded down, but the wolves themselves were nowhere to be seen.

Normally Jim would be busy getting a fire started in the woodstove, but he could tell something was different that morning too. He was already dressed for the cold and was assembling his camera gear. We didn't have to guess what was going on; we'd been preparing for this day for the last two months, and we knew the wolves have been anticipating it just as much.

We pulled on our boots. There was no need for snowshoes, as the cold night had frozen the snow to a hard crust. We headed uphill, past the meadow, to where bare willows gave way to a grove of mature spruce trees. There we found the pack in a mass of energy, gathered around a hollow at the base of a fallen tree. They shifted left and right almost as if they were dancing on tiptoe. With their noses thrust forward they sniffed the air and whined to each other in excitement, pausing every so often to cock their heads and listen. We didn't want to interfere with their moment, so we kept our distance, but every so often we could hear a sound above the commotion: a faint birdlike chirp rising out of the dark hollow beneath the tree.

Chemukh was the slender black female who had joined the pack as a pup two years earlier. When she reached maturity in February, she and Kamots had mated, making her the alpha—the pack's only breeding female. In the early weeks of April this nervous female took on a new earnest attitude. She'd often break away from the others and spent her days searching for the perfect location to give birth. She'd pick a new spot, dig a foot or so into the earth, then decide that the soil was too hard, too damp, or too something, and she'd move on. Her intensity was contagious, and when the other wolves saw her at work, they began to dig too. They weren't helping; they were just caught up in the spirit and dug random holes here and there. At last Chemukh settled on the fallen spruce and began working away at the hollow where the roots had pulled up. Once she had chosen that spot, it only took her a few days to excavate a den to her satisfaction.

Every wolf seemed to be aware of the importance of her work, but none more than Lakota. The pack omega, he often bore the brunt of Chemukh's hot temper, but instead of avoiding her, he started parking himself beside the den almost as if he were standing guard. It was just one of the many indications of how much the entire pack was wrapped up in this event. At last, early on that April morning, Chemukh had entered the den alone and given birth.

When we arrived at the fallen spruce, Jim cautiously set

up his camera some 20 yards away. It was a highly charged moment, and as eager as we were to document it, we weren't about to risk disturbing the new mother. Over the next hour or so we inched, step by step, tree by tree, closer to the den. Every so often Chemukh popped out of the hole to be greeted with a flurry of licks from Kamots and her packmates. Then she would disappear underground again. It was almost as if she were giving progress reports, letting the others know all was well. She was aware of Jim and me standing close by and didn't seem to mind our presence at all.

The pack's energy was intense, and I began to record their sounds, trying to capture the excited conversation that the wolves were having. After a couple of hours I put down my gear. Jim and I had already discussed how we would handle this moment. From the time she was born, I had developed the closest bond with Chemukh, and she was relaxed and trusting around me. We had therefore concluded that I should be the one to approach the den and check on the newborn pups.

Slowly and deliberately, I stepped forward to join the seven wolves gathered around the den. They knew me intimately; I had been a constant presence in their lives for years. All the same, I never wanted to take their trust for granted. Amped up as they were, I wanted to present myself as calmly and respectfully as possible. I especially wanted to be careful around Chemukh. She could be skittish, and I didn't know how defensive she would be around her pups. So as I leaned forward to peer into the darkness of the den, my heart nearly skipped a beat when I found myself staring straight into the yellow eyes of the mother wolf.

To my relief, Chemukh's eyes showed no fear, only curiosity. I made way for her, and she crawled out of the den and stood beside me, regarding me, perhaps trying to read my intentions. I spoke softly to her, telling her I wanted to have a quick look at her pups. I had a little flashlight with me, and I calmly showed it to her, letting her sniff it to assure her it wasn't anything dangerous. Then I slowly lowered myself onto my belly, flicked on the light, and eased my way into the den.

Chemukh had never dug a den before, nor had she watched a parent dig one, yet her work was masterful. Aboveground snow was melting into a muddy slush, but the interior of the den was dry as a bone. She had hollowed out a tunnel about six feet long, ending in a wide chamber. At the very back, she had carved a small shelf to keep her pups high and dry. Pointing in that direction, my flashlight beam caught a squirming mass of dark fur. The pups' eyes were still shut, but they could sense my presence. They lifted their heads and whined softly, probably hoping I was there to feed them.

There, in the darkness of the den, I felt time slow down. Suddenly I was keenly aware that this was a pivotal moment in my life. I felt completely at peace and fully in the present, as though all other phases of my life had been mere preparations for where I now found myself. Nine years earlier I had struck up a chance conversation with a man on an airplane. A hundred twists and turns later, and there I was in the middle of wilderness, underground, sharing an intimate life event with a pack of wolves—and everything about it felt exactly as it should be.

I didn't want to overstay my welcome, so I wiggled my way backward toward daylight. Chemukh was right there looking at me as I wiped the mud from my clothes. I congratulated her, and she cocked her head in response. Then she delicately leaned forward and gave me a little lick on the nose, then turned and disappeared back into her den. Until that moment my eyes had been dry, but I felt myself welling up at last. It was a gesture of complete trust that I will never forget.

That was my first glimpse of the trio of black pups we would come to call Ayet, Piyip, and Motaki; we named one after the previous omega who had been killed by a cougar. They were the fourth and final litter that we would raise at wolf camp. More significantly, they were the first wolf pups born in the Sawtooth Mountains in over 50 years. To us, this moment was nothing short of historic.

Of course, the wolves were completely unaware of the broader significance of this occasion. Their joy was something much more fundamental and pure: Their tribe had just received brand-new members; their extended family was growing in number and in strength. Every wolf in the pack shared in the celebration. For days after Chemukh gave birth, they paid visits to the den, crowding around to sniff and listen and peer inside. Then they'd go off together, shoulder to shoulder, vocalizing excitedly in their "Chewbacca voices." How I would love to know what they were saying to one another, but I think I got the gist of it. The pups may have been born to Kamots and Chemukh, but they were precious to every wolf in the pack.

WE KNEW GOING INTO THIS PROJECT THAT WOLVES were social creatures, but after living with them, we have come to understand their bonds as something even deeper. Each wolf knows its family members intimately and cares for them deeply. When their family is healthy and strong, every wolf feels and shares that sense of strength. Thus, every wolf has a stake in caring for the entire family, especially new pups. Family is simply everything.

Wyakin, the other female born at the same time as Chemukh, was the first to step in to help raise the pups. When Chemukh needed to take a break from the responsibility of motherhood or join the others on an elk carcass, Wyakin would immediately fill the void, curling up with the litter so they'd always have a warm place to cuddle. There were days when Wyakin appeared to spend as much time with the pups as their mother did.

We knew we would soon have to socialize the pups so they would come to trust us just as the adults did. We had learned from Erik Zimen and others that even though the adults trusted us, they would still teach the pups to be fearful of humans. The last thing we wanted was for any of the wolves to feel stressed by our presence. To make sure Ayet, Motaki, and Piyip felt bonded to us, we chose to hand-raise them for several weeks. That meant round-the-clock care, lack of sleep, and bottle-feeding little creatures with needle-sharp teeth and claws, which left us bloody but happy. It was the proven technique we had used on all the other wolves in the pack and the reason all the wolves felt safe and relaxed in our presence. When we returned the three pups to their family, the wolves were thrilled to have them back. They immediately picked up where they left off.

At nine weeks the pups were already eating solid food. Mealtime was a time when the pack hierarchy came into play, but the pups were not part of the adult hierarchy just yet. They were allowed to join the meal whenever they pleased and eat as much as they could. We watched with fascination, as we had with previous litters, to see the reaction of the adults the first time we brought in a deer carcass for them and the pups. Motomo growled at Amani, Amani growled at Lakota, and Lakota submitted to the others. The pups, however, dived headfirst into the meal with total abandon, and the adults happily stepped aside to make room for them.

No matter how much the pups ate, they always had room for more, and Matsi was always eager to accommodate them in his remarkable way. As the pack's beta wolf, he never had to worry about getting pushed off a kill and was always able to gorge himself to the point of nearly bursting. After eating, he walked a few yards away. Piyip, Ayet, and Motaki took the cue and bounded after him, crowding around him and licking his muzzle. For Matsi, this was an instinctive trigger. He promptly regurgitated a large portion of the meal he had just eaten, and the pups eagerly lapped up their second course.

I admit that I was never able to watch this without my stomach turning a little, but this behavior, an instinct born of necessity, works very well for wolves. They have no other way to bring large amounts of easily digestible food from their kills to pups waiting several miles away at a den or rendezvous site, so they fill their bellies with meat and, when they see the pups again, up comes a meal. A bit of predigestion is just the thing for pups getting started on solid food. During the pups' first summer we saw every wolf in the pack, even Lakota the omega, offer up food in this way, but Matsi was the most consistent provider.

Matsi's attention to the pups was gentle but earnest. He would join them in play, but his main concern was their safety and security. While the pups tumbled over each other, growling and mock fighting or playing tug-of-war with a piece of hide, Matsi stood by as their faithful protector and pup-sitter. Piyip was curious and bold like his father Kamots. He'd occasionally get bored playing with his sisters and try to wander off, but Matsi would sidestep and corral him back into the group.

Although Matsi was always looking out for the pups' safety, the mid-ranking wolf, Amani, only seemed concerned with their amusement. Amani was an interesting wolf, to say the least. He embodied how complex wolf personalities can be, and how different one individual is from another. He was often quite domineering, but around the pups he transformed into a goofy uncle. I remember watching the pups playing when Amani meandered up and flopped down in the middle of the pile. Piyip immediately grabbed his tail and started pulling, growling his fiercest little pup growl. Ayet started gnawing on his ear and Motaki seized his hind leg, shaking it as if she'd just caught a bull moose. Amani rolled over onto his back as though he were getting a massage. His jaw slackened into a half smile as his eyes fluttered shut. He was in heaven.

Years later, while watching wolves in Yellowstone, Jim and I were lucky enough to meet Nell and Bob Harvey, a pair of dedicated wolf spotters. Many people think that professional biologists and park rangers are the only ones conducting wolf research in Yellowstone, but actually people like Nell and Bob also help out. For seven years they've been coming from California to volunteer their time. When we met them in 2016, they were well into their annual total of 111 days observing and recording the behavior of the Yellowstone wolves. At 8,000 feet above sea level, in brutal sun or savage cold, they patiently watched, noted, and reported their observations for no other reward than the joy of watching wolves. These animals can do that to people.

One of our favorite stories Bob and Nell ever told us involved a young male that we imagine may have been quite like our Amani in his inclination to spoil the young pups of the pack like an indulgent uncle. It was early spring, and this yearling from the Junction Butte Pack was wandering about, probably hunting pocket gophers. Always playfully interacting with his younger packmates, as our Amani did, the young wolf seems to have decided to bring them a gift. At this time of year the pups born to the alpha male and female had moved out of the den but weren't quite ready to follow the pack on long journeys, and so they were left under the supervision of another pack member, a pup-sitter, at a rendezvous site—a safe area that serves as a base camp until the pups have matured.

On the way back to the rendezvous site, this particular young male found the perfect gift: the skull of a dried-out bison carcass. There wasn't any food value in the skull; it was essentially a big toy. As Nell and Bob watched with spotting scopes, the wolf struggled with this unwieldy present, climbing for more than a mile through rolling sage hills and over rocky ridges. Normally a wolf could cover this ground in a matter of minutes at a steady lope, but his prize was so heavy and awkward that it took him over an hour. He had to stop several times to rest. At last he delivered his precious gift to his young brothers and sisters, who commenced to gnaw on it and play a game of "keep-away," as each one tried to drag the heavy object away from the others.

As we listened to this story, I thought about how this dedicated young male must have felt as he watched the pups enjoying his present. He had discovered this fantastic plaything and decided his young packmates absolutely had to have it. Nothing would deter him from delivering it to them.

What could be behind this total devotion toward the pups? I've often wondered. I suppose one could speculate that Matsi and Amani were trying to form alliances with young impressionable pack members, but I don't really believe a wolf could be that calculating. Wolves are far too transparent. They bare their inner lives moment to moment in their body language and vocalizations. When I watched them interact with the pups, all I saw was unclouded affection and care. This was their pack—their family—and family meant everything to them.

Family certainly meant everything to the Junction Butte Pack as well, but the pack was beset by a chain of calamities. By the summer of 2016 the family was grappling with the death of the alpha female and her new litter. In an unusual turn of events, earlier that spring two subordinate females also mated and produced nine pups between them. In the absence of the alpha female, the pack's yearlings were stepping up their game, helping to raise new pups. Tragedy struck yet again when the alpha male died in early September.

Then, on September 24, the pack crossed the park boundary and ventured into parcels of land known as "Wolf Management Units." In these areas, wolves can be legally killed—or, to use the cheerful official term, *harvested*. Wolf hunters wait there expressly for the chance to *harvest* Yellowstone wolves the moment they step outside the federal protection of the park. On this day hunters killed one pup and two yearlings— a female and a male. The yearlings hadn't yet been tagged or given official identification numbers, but their markings were distinctive. The dead male was certainly the exuberant youngster who Bob and Nell had watched as he proudly delivered a gift to his young brothers and sisters.

Since then, two more Junction Butte pups and another yearling have disappeared, but officials do not know if they are living or dead. The human assault on the Junction Butte Pack compounded the indiscriminate blows that nature delivered. The pack fragmented and lost territory, and for a while it seemed it would collapse under the strain of these tragedies. But wolves are resilient. The pack recently added new members and has begun slowly to recover. All the same, we'll never get to see what kind of wolf that yearling would have become.

WHAT IS A WOLF FAMILY? Anthropologists suggest that wolves organize themselves similarly to the way early human societies did: as nuclear families within extended families. In our two species, the basic family unit begins the same way: Boy meets girl, they form a pair bond, settle into a home, and have offspring. After that, anything can and does happen. Grandparents stick around, aunts and uncles move in, children leave and may come back, and—occasionally—complete strangers are welcomed into the fold and treated as family members. If territory is available and prey is abundant, alphas may permit other members of the pack to breed. One unusual year in Yellowstone, the alpha male of the famous Druid Peak Pack mated with three females. Rules, it seems, are meant to be broken. Young wolves born to neighboring packs also worked their way into this Yellowstone pack that year. The result was one great big extended family of 37 wolves. It's safe to say that once a wolf pack grows beyond the simple nuclear family, anything goes.

Wolves, like people, are predisposed against inbreeding, so welcoming new blood is necessary if a pack is going to survive for multiple generations. That makes practical sense, but wolf packs do something else that can't be explained as easily. When wolves encounter pups—related or not—they frequently adopt them. It just seems to be an instinctive and irresistible desire.

People have observed this behavior in wolf packs all over the world. Our friend the late Gordon Haber, who studied the wolves in and around Alaska's Denali National Park and Preserve, focused much of his work on a group of wolves who lived along the Toklat River. Over his years of study he watched wolves take just about every conceivable approach to family life.

During a two-year period we followed Gordon all over the state of Alaska, in backcountry planes equipped with skids to land on snow, trekking through wilderness, crossing waist-deep roaring rivers, and encountering plenty of grizzly bears and relentless mosquitoes. From a small hill overlooking the Toklat River we spent part of a summer observing a pack that Gordon had been studying. The wolves were much too far away to film, but his stories about their lives were captivating. Gordon emphasized how important family was. He explained that wolves pass knowledge on to future generations, teaching their offspring where and how to hunt, where to cross a river, and how to survive. His observations were rich with personal stories of wolves he knew and, as one could see, admired and loved. Some of the tales were even soap opera–worthy, such as the one involving a mother wolf who left her mate, took the kids, and moved back home to live with her mother. Other accounts were quite touching.

In March 2001, the Toklat's alpha male died during a radio-collaring accident, leaving his pregnant mate and seven older offspring. This is Haber's account:

> Two adult sibling males from two hundred miles away showed up at the natal den in early June shortly after the female had produced her dead mate's new pups. One of them soon became the new Toklat alpha male, and both helped raise the unrelated dead male's pups without any obvious difference in effort or affection between the mother and her seven older offspring.

Was this particular male wolf just especially tolerant? It doesn't seem like it.

In another ongoing study, Doug Smith has been observing the most famous population of wolves in the lower 48 since 1995. Doug is the project leader for the Yellowstone Wolf Restoration Project. He's the point man on all the wolf research carried out in the park, and runs the whole operation out of a small office in the Mammoth Ranger Station.

Competition between packs is high in Yellowstone, where many wolves now live in close confines. When conflict erupts and packs do battle, young pups can become casualties just as adults can. But when a revolution in pack leadership takes place, it's a different story. When an outsider, or an alliance of outsiders, unseats one of the alphas in an existing pack, the victors always treat the pups of their former rivals with parental tenderness. This isn't true of many other animals. Male lions are especially notorious for infanticide. When a usurper takes over a pride, his first act is often to kill the cubs of the former male. Even some of our primate relatives will kill unrelated offspring after a transition of social power, but not wolves.

Doug generally maintains a healthy scientific distance, but even he marvels at this special quality. "In lions," he says, "the male kills all the cubs instantly, but wolves don't do that, and we've seen this time and time again. Males join packs and care for pups that aren't related to them. Evolutionarily it doesn't make sense, but every time it happens. Every time. They don't kill the pups."

Another story about nurturing stepfathers has been observed in Yellowstone's Hayden Valley. Three outside males had taken over a pack and supplanted the alpha male. They didn't hurt him; they just pushed him out. Now he's lingering on the fringes of his old territory while his former mate and three newcomers raise his pups. When the adults are off hunting, the ousted male comes in to visit his kids, but the new wolves do most of the work of feeding and caring for them.

Anyone who has been to a dog shelter knows that irresistible wave of protective instinct that strikes the moment one looks down at a helpless little puppy. As human beings we're programmed to respond to certain traits—big round eyes, soft features—with protective and nurturing behavior. Even people who claim they are not particularly crazy about children couldn't walk by an abandoned infant. It's not within our control: We're hardwired to protect infants. I believe this is something wolves share with us.

EARLIER IN OUR PROJECT, two years before Chemukh dug her den under the spruce tree, there was a time when she, too, was a little pup, brand new to the pack. But she wasn't born to them. We introduced her to the Sawtooth Pack in 1994 along with her siblings, Wahots and Wyakin. At that time what we called the Sawtooth Pack was a group of five males—no females, hence no breeding pair. We knew that if these wolves were to become a true pack, we had to introduce some females, and that was the impulse that brought Chemukh and her sister and brother into the family. We raised them by hand, as we had learned to do with the previous two litters, to make sure they would be socialized to us and regard us without fear. By that time our camp was an enclosed little island in the middle of the pack's territory, so we could keep the pups with us around the clock while allowing the adults to get used to the pups' presence and vice versa through the camp fence.

The five adults were completely enraptured by the sight and sounds of the young pups. Their favorite pastime quickly became gazing at Wahots, Wyakin, and Chemukh as the pups frolicked around the tents. When the pack would rally together for a howl, it was absolutely endearing to watch the three pups try to howl back in their awkward, exuberant, yodeling way. Kamots and Matsi even brought bones and pushed them through the chain-link as little welcoming favors.

Once we were sure the pups had bonded to us and trust was established, it was time to introduce them to the rest of the pack. We were all a bit nervous, I think—the pups, the adult wolves, Jim, and me. At last we swung open the gate and let the wolves take it from there. We expected that the pups would charge through to join the adults, but it turned out they weren't as ready for this moment as the adults were. Kamots sensed their apprehension. He calmly entered our camp and walked over and gave each pup a gentle lick as an introduction. I was recording sounds and listening to their elaborate conversation through my headphones. It was an emotional dialogue of whines, yips, growls, and the ever-endearing Chewbacca noises. The body language of joy mirrored their excited vocalizations. Tails wagging broadly, each adult introduced itself to the equally animated pups. Amazingly, the pups knew the protocol and instinctively rolled onto their backs, exposing their vulnerable bellies in complete submission. Through it all I inferred that the big wolves were offering the little ones a bit of reassurance but at the same time explaining their hierarchy. Every sound, every posture seemed to say, "You're going to be okay. You'll be part of our family now. Everything is going to be fine."

The little threesome took a bit of convincing. The sheer size of the adults must have been intimidating. Chemukh, always the most skittish of the siblings, ran and hid in the willows for an hour or so while the frustrated adults tried to coax her out. Wyakin, the little gray female, immediately found a mentor in Matsi and clung to him for courage. Her brother, Wahots, tried to stick to his sister's side, but the whole meet and greet became too much for him and he disappeared. We figured he was hiding in the willows and would come out eventually. After recording sound for about an hour, I went back to our tent to stow my gear. As I sat, quietly logging my tapes, I got the uneasy feeling that I was being watched. I turned around and there was little Wahots, poking his head out from under our cot. Somehow, with all the comings and goings of the wolves through our camp gate, little Wahots had managed to sneak back into a place where he felt safe. Now he was looking at me with a bewildered expression as if to say, "What am I supposed to do now?" I picked him up and guided him back through the gate, feeling like a mother dropping her child off at the bus stop on the first day at school. Matsi was waiting there to greet him with little Wyakin in tow. The big wolf greeted Wahots gently to let him know that he was safe and welcome. By evening they were heading off into the meadow together toward the forest, with puppy-sitter Matsi walking calmly ahead and Wahots, Wyakin, and Chemukh bounding playfully behind.

As we settled down to dinner in our cozy yurt that evening, we couldn't stop talking about how vocal the wolves had been as they were greeting the pups. In all the things I had read about wolves, I didn't recall reading anything about how varied and musical their conversations were.

When people talk about the language of wolves, they usually focus on howling, but there's so much more. Pack rallies culminate in a howl, but before and after that climax, we've always heard the wolves making lively conversation—and it's very clear to me that their vocalizations are expressions of family solidarity. We'd hear Kamots talking discreetly to his timid brother when Lakota needed encouragement. Wyakin and Wahots, inseparable siblings, would converse with each other as they'd gang up on their littermate, Chemukh. Amani and

Motomo would sometimes work together to outsmart their leader Kamots and snatch extra food, and we could hear them hatching plans in their quiet voices before springing into action. It has occurred to me that most wolf observations take place at a distance that is too far away to hear this subtle language. Only because the Sawtooth Pack awarded us their trust and allowed us to live within their territory were we able to hear them. It proved to me that wolves are interacting all the time, touching base with each other, communicating moods and reaffirming their family bonds.

The years after Wahots, Wyakin, and Chemukh joined the pack were the richest years of our project. We watched as the Sawtooth Pack grew from a mellow clan of five males, to a proper pack of six males and two females, and at last to a dynamic family of 11. Whether it was their adoption of Wahots, Wyakin, and Chemukh or the arrival of their own litter of Ayet, Motaki, and Piyip two years later, the Sawtooth wolves rejoiced when their family grew. That is not to say there were not family squabbles. Lakota may have had to endure the omega status at the bottom of the hierarchy for many years, but he was as much a cherished member of the family as any of the others. He always ate, he howled with the group, he played with his packmates, and he helped raise the pups.

We have marveled at each individual wolf's intelligence and been fascinated by observing each unique personality, but the family bond shared among these wolves is what we have admired the most. Caring for the young ones—and for each other—was the central mission in their lives. A wolf is impelled by many individual desires—it wants to breed, hunt, perhaps explore—but its most profound desire is the one that touches us at our very core as human beings: A wolf wants to belong.

CHAPTER THREE

LEAD WITH KINDNESS

Jim

THE ASPEN TREES WERE SURRENDERING after their brief moment of autumn brilliance. Their spade-shaped leaves had faded to dull yellow and brown and began to shake loose with each gentle puff of wind. One by one, ravens gathered in the barest branches. The birds engaged in a lively conversation, chatting back and forth in a musical gurgle that always sounded to me like a strange mechanical dribbling of water. They seemed to enjoy pausing to hear the sound bounce back to them from the mountain walls. One bird swooped down to join three others picking at some scattered bones and pieces of hide—the signs of a wolf feast completed.

The wolves of the Sawtooth Pack paid the ravens no mind. They had spent the last three days working through the carcass of a 600-pound bull elk and were enjoying the resulting food stupor, resting at the edge of the forest on a

Kamots, the alpha or pack leader

crisp October afternoon. During our years at wolf camp, we did our best to re-create the way wolves live in the wild, mimicking the success or failure of a hunt, bringing in road-killed elk and deer at irregular intervals. After such a big meal, they wouldn't need to eat again for several days.

It was a picture of pure contentment. Motomo and Amani basked in a patch of sunshine. Snarling and growling in mock battle, Wahots and Wyakin played tug-of-war with a leftover piece of elk hide. Matsi and Lakota snoozed side by side. Chemukh, always the loner, gnawed on the end of an antler a few yards away. Kamots sat and calmly watched his pack.

Suddenly he lifted his head to full attention and his body became tense. The forest was quiet save for the occasional croaking of ravens, but his ears pivoted forward and he turned to point his nose toward something distant and unseen. With sudden purpose he rose to his feet and trotted away with that cool, determined gait that wolves have. Motomo and Amani lifted their heads to watch him go, and then settled back down. They knew Kamots. Whatever it was, he had it covered.

Fifteen minutes later he reappeared at the same confident trot, rejoined the others, and lay back down. It must have been a false alarm, possibly an elk moving through the nearby forest or the distant falling of a long-dead tree. His body language told all that there was no cause for concern. Whatever was going on, he dealt with it. That's just who he was.

For the better part of nine years Kamots was the undisputed leader of the Sawtooth Pack. Under his protection, the pack thrived. From a filmmaker's perspective, he was beautiful to behold. His looks were classic—gray with a charcoal saddle, a dark face mask above a light muzzle. He wasn't the biggest wolf in the pack, but the way he stood often made him appear so. He held his head and tail high and kept his ears up and alert. When he moved, it was with an unhurried confidence and certainty. His eyes were his most expressive feature. They were light amber, almost yellow, and in an instant they could flash from serious to mischievous, to concerned, to disarmingly sweet. The joys and the burdens of his role seemed to play across his face constantly, making him one of the most engaging wolves to film, photograph, or just watch. In looks, in bearing, and in behavior he was the embodiment of an alpha wolf.

We've noticed lately that the word "alpha" is falling out of favor among some biologists. They would rather use the dry observational term "breeding adult." Unfortunately, the word "alpha" has also come to be used to describe aggressive, hypercompetitive human males, which further conveys the wrong impression. Yes, alphas are the breeding male and female—but they are so much more than the dominant individuals of the pack. After spending years in Kamots's company, we have concluded that being an alpha has almost nothing to do with aggression and everything to do with responsibility. Alphas are driven from within to shoulder the well-being of the entire pack. They patrol the boundaries of their territory, looking for danger. They are keepers of pack knowledge—where to find prey and how best to hunt it. Alphas are assured, alert, and compassionate. A true alpha is a leader in the very best sense.

KAMOTS WAS BORN IN THE SPRING OF 1991. He and his littermates were the first pups we raised by hand to create the foundation of the Sawtooth Pack. From my very first experiences with Kamots as a pup, I could tell he was special. Some wolf pups are playful and submissive, some are nervous and shy, but Kamots was bold and curious from the start.

When they were two weeks old, we introduced these three pups to a fenced-off area outside where they could get used to the natural world. At first they huddled close together and timidly sniffed the grass, but after just a few seconds little Kamots bounded forward, leaving his siblings behind to inspect every corner of his new surroundings.

The social structure of a wolf pack is hierarchical, and pups develop their own "pup hierarchy" right from the beginning. It was amazing to witness the difference between Kamots and his brother and sister, even at this early age. When we had to separate them for bottle-feeding, Lakota and Motaki would become quite anxious at being apart from their siblings. Then, when they were reunited with the group, there would follow a period of squabbling, almost as though they had to reestablish their hierarchy all over again. Kamots, on the other hand, took being alone as just another experience. When he rejoined his littermates, he reassumed his place calmly, as though he had never been gone. As Kamots grew to adulthood, he simply assumed the role of alpha male as though it had always been his. He never fought another wolf for the position. He never dominated another wolf with excessive force or intimidation. He simply took charge. When we introduced the pups Matsi, Motomo, and Amani the following year, they accepted him as their leader without any hesitation.

One of the things that Jamie and I enjoyed most during our years with the pack was to wake before dawn and watch from our camp as the wolves said good morning to their fellow packmates. They didn't sleep solid hours like we did. Each wolf had a few favorite places to sleep, and they'd move about during the night. Still, to our amazement, each wolf began the day by paying respects to Kamots, and each in its own unique way. Matsi was the pack's beta wolf, second only to Kamots in rank. He possessed a similar confidence but he was mild-mannered and content to let someone else lead. When he greeted Kamots, Matsi would often hold his head slightly lower than the alpha's and stand beside him like a loyal lieutenant. Lakota, the lowest wolf in the hierarchy, would practically belly-crawl toward his brother. He'd lick Kamots's muzzle the way a pup begs for food, whining softly. There was no fear in this ritual. It seemed to be a way of showing the leader respect. For his part, Kamots's demeanor was not one of severity but one of care and concern. He was making sure all were present and accounted for and everyone was well. Above all, the wolves were expressing the bond that each felt for the other, and all recognized Kamots as the center of their universe. They greeted each other, touched noses, and pressed shoulder to shoulder in gestures of camaraderie with Kamots as the hub.

As the protector of the pack, Kamots was responsible for keeping order. He was usually a calm, cool presence, but there was no getting around the fact that he was also the head of the hierarchy. From time to time he had to assert himself and make that clear, most often when a deer or elk carcass was present.

I'll never forget watching him over a meal when he was only about a year old. We'd found a small deer that had been killed on the highway and brought it to feed the pack. There wasn't a whole lot to go around. As Kamots stationed himself over the kill, his brother Lakota approached timidly, hoping to grab a mouthful. At first Kamots simply raised his eyes and stared at his brother, uttering a long, low growl. Lakota shrunk lower, trying to look as insignificant as possible, but he couldn't help stretching his neck forward toward the deer, just trying to get that one bite. Kamots's face transformed. His lips curled upward to expose a mouthful of teeth and his ears shot out to the side like horns. He let out an assertive growl and lunged. Lakota leaped backward with a yelp and flipped over in submission. The episode was surprising in its sudden ferocity, but I came to learn that this is the language of wolves. It was not Kamots's intention to keep Lakota from eating entirely—just to make him back off for a bit and respect the pack's social hierarchy. He made his point without ever making contact. By the time Lakota crept toward the kill once more, Kamots had settled, and he allowed his brother to eat.

Kamots's leadership was firm at times, but we never saw him treat any wolf viciously or harm any member of his pack—this was not always true of the other wolves. Amani, despite showing endless patience with young pups, could be a bully toward other adults, especially Lakota. I suspect it was because he was afraid he might drop in rank if he didn't continuously prove his strength. Kamots, in contrast, never felt the need to behave this way. His authority was secure. A warning growl and a flash of teeth were sometimes necessary to keep everyone in line, but he almost never took it further. I believe that he earned the pack's respect in a way that a more aggressive wolf would not.

By the time the three pups, Chemukh, Wyakin, and Wahots, had made it through their first summer, Kamots began to treat them with a slightly firmer hand. For their first four months of life he had guaranteed them a place at every carcass and made sure they ate their fill. Around September, though—when the pups were adolescents—he changed his tune. He began challenging them and making them assert themselves for the right to eat. I believe he was doing some character building, because he focused on timid Chemukh especially. Before they're a year old, young wolves must be ready to travel with the pack, making their way through deep snow and participating in the hunt when they can. I think Kamots was letting the growing pups know that childhood was over and it was time to toughen up.

KAMOTS WAS A MARVELOUS LEADER, but he was not the Sawtooth Pack's only alpha. A wolf pack actually has two alphas and two parallel hierarchies—one for males and one for females. In the final years of the project, Chemukh emerged as the Sawtooth Pack's alpha female. Although Kamots was the very embodiment of a natural leader, Chemukh was anything but. She was high-strung and could be aggressive when she felt insecure. She was, however, one of only two females in the pack, and her counterpart, Wyakin, was easygoing and not at all prone to fighting. When the two were still immature, Jamie and I speculated that Wyakin's even temperament would make her the better alpha, but biology had other plans. Chemukh came into estrus—her first period of sexual receptivity—before Wyakin, and Kamots took notice at once. By the time Wyakin caught up, Kamots and Chemukh had already mated, securing Chemukh's position. From that point forward she became something of a tyrant, going as far as biting Wyakin and Lakota from time to time. As much as Kamots showed me what a leader was, Chemukh sometimes showed me what a leader wasn't.

When I think of Chemukh, I'm reminded of a story from Yellowstone's famous Druid Peak Pack and a wolf who is fondly remembered as the Cinderella Wolf. Cinderella—or 42, as her radio collar officially identified her—was a mild-mannered wolf, subordinate to her sister, number 40. Unfortunately, that alpha female, number 40, reminds us of Chemukh, not the even-tempered Cinderella. Wolf number 40 was a textbook tyrant who maintained her status through a steady outpouring of aggression toward the pack's other females.

After decades without wolves, Yellowstone was flush with an overabundance of elk. Under such favorable conditions, a wolf pack may allow multiple members to breed. In 1999 the Druid Pack's alpha male sired pups with both 40 and 42 (Cinderella). This seemed to push 40, the alpha female, over the edge, and Cinderella found herself the target of her sister's wrath more than ever. At every opportunity, 40 attacked Cinderella and attempted to drive her out of the pack. She may have even killed Cinderella's first litter, though no one can say for sure.

Wolves generally cooperate in all aspects of family life, but the following year, when Cinderella again mated with the alpha male, she was so concerned for her pups' safety that she dug a den far from her sister's and kept her young ones out of sight. Then one day, 40 came calling. The aggressive alpha attacked her sister as always and threatened her pups. At last Cinderella turned and stood her ground. At this point, the rivalry between the two females could have gone either way, but something truly amazing happened: The other females in the pack rose up and rallied around Cinderella. Together they literally staged a coup, killing the tyrant and elevating a new leader, one they knew would be even-tempered and kind. From then on, 42—the Cinderella Wolf—ruled benevolently as the Druid Peak Pack's alpha female.

Fortunately, as Chemukh aged and settled in as the Sawtooth Pack's female leader, she became more secure and her aggression subsided. She was never to be the strong leader that Kamots was, but at least she became a reliable member of the pack.

Lately biologists have been trying to determine whether the alpha male or female is the ultimate "leader of the pack." Doug Smith, the head of the Yellowstone Wolf Restoration Project, has spent thousands of hours observing the park's wolves, and even he is unable to give a definitive answer to this question. He did share some stories with us, though. It's quite common, he says, to see an alpha male stand up and stretch, looking like he's ready to venture off somewhere, and no other members of his pack pay much heed. On the other hand, very often when the alpha female stands up, every other wolf will come to attention, as if to say, "Oh, we're doing something now." It doesn't happen every time, Doug says, but his observations suggest that wolves often appear more keyed in to the alpha female's leadership than to the alpha male's.

It could be that in the most successful packs the alpha male and female rule together, dividing their duties based on their individual talents. Big males are good at guarding territory and protecting the pack. On the hunt, the alpha male is often the one to deliver the finishing blow to prey. On the other hand, females are usually smaller and faster on their feet. They can often be seen initiating the hunt and testing a prey animal to see if it's worth pursuing. Jamie and I noticed another sign of the alpha female's elevated status: When Chemukh dug her den, no other wolf, not even Kamots, entered it. That was her domain and hers alone. Because a wolf pack is above all else a family and raising healthy pups is the pack's primary mission, it's not surprising that the mother of the pups would play the key role.

Still, I can say without question that in the Sawtooth Pack, Kamots was the one in charge. He alone patrolled their territory and kept an eye out for danger. When they

gathered for a pack rally—a spirited display of pack solidarity—the other wolves rallied around him. When they
howled, Kamots led them in their song. Whether the male
or female—or both—the alpha is the center of gravity that
holds the pack together.

After many years leading her pack, the Cinderella Wolf
was found dead near an elk carcass at the edge of her territory. A few months later, her mate, a grizzled male called
21, left the pack and walked alone to a ridge, laid down
under a tree, and died peacefully. It was a momentous time,
reshaping the world of the Yellowstone wolves. Imagine the
sun abruptly vanishing and the planets, with no force to
hold them in orbit, spinning off into the void. For the
remainder of the Druid Peak Pack, that's how it probably
felt. Some splintered off on their own to seek out new territory. A small pack under a new alpha pair tried to rebuild
on the old territory, but the pack never regained its former
power. By 2010 the Druid Peak Pack had ceased to be.

Wolf packs aren't made to last forever. During their long
reign, Cinderella and her mate 21 likely raised more offspring than any other wolf pair in Yellowstone. Their
blood runs in many of the wolves who lead packs today
throughout the greater Yellowstone ecosystem. So
although it is sad to think of the passing of these two well-
known wolves of Yellowstone, it is worth remembering
that they were victorious. The greater tragedy is that many
great alpha wolves out there today do not get the chance
to have that success.

We know from our own human experience that stories of heroism are often tragic tales. The courage, compassion, and sacrifice that make a human—or a wolf—a great leader often sow the seeds of their downfall. Consider these facts: In Alaska trappers run their lines right along the boundaries of Denali National Park and Preserve, taking advantage if any wolf steps across the invisible border. In Yellowstone and Glacier National Parks, wolf hunters may even be using radio telemetry to track the movements of radio-collared park wolves. If any wolf leaves the protected area (and in the pursuit of migrating elk, they inevitably do), hunters are already lined up and waiting. In the Idaho wilderness, there is no safe zone for wolves, and hunting them recreationally is encouraged. In 2014, Idaho's Department of Fish and Game sold 43,300 wolf hunting tags, or permits. That year there were a mere 650 wolves in Idaho. In several regions in Idaho today, including in federally protected wilderness areas, management plans in place are designed to reduce already tiny populations of wolves by as much as 40 to 60 percent. In such a hostile environment, what would have befallen a leader such as Kamots? He was always the one to investigate danger, and he would never have hesitated to put himself between a threat and his pack. It's what great leaders do.

So when hunters are combing the wilderness looking for wolves, which ones are they going to see—and shoot? Every time we read a headline about wolves getting shot, we have come to expect that at least one of an alpha pair was among the victims. Protecting the pack is the alpha's solemn duty—and may likely be his or her doom.

When an alpha is killed, the pack goes into shock and grief. We're talking about young and mid-ranking wolves literally losing their mother or father. Wolves suffer this emotional loss intensely. The pack has just lost experience, knowledge—and its gravitational center. Just as we saw when the Druid Peak Pack lost its alphas, the surviving wolves often scatter. A fragmented and traumatized pack of two or three wolves without an alpha will have a much more difficult time bringing down an adult elk or even defending their kill from bears and other predators. Ranchers may even suffer from the impact, for a smaller pack of less experienced wolves may go after easier prey, such as livestock.

A large pack with strong leaders is far more predictable. The alphas maintain their territory so they can hunt on familiar ground. Once alpha females establish a safe den site, they'll often return to it year after year and even pass it on to the next generation. Experienced alpha wolves know the danger that humans present. Knowledgeable alphas will pick a safe place to hunt and raise young that minimizes conflict with rival packs and with humans—and once they've taken possession of that locale, they will hold onto it.

When we talk to policy makers, we often stress these practical issues because of their repercussions through the landscape that wolves share with humans. It is easy to discuss the importance of an alpha as the one that brings experience, strength, and stability to a pack. Kamots certainly had those attributes. But for Jamie and me, there was something more difficult to define—something in the alpha's mere presence. I may have been the human in charge of the Sawtooth wolf project, responsible for all logistical and practical matters, but I looked upon Kamots as the true leader and spirit of the pack and the project.

By the time our wolf camp project reached its halfway point, the anti-wolf community was making it far more difficult and disturbing than I had ever imagined it could. One autumn day in the project's fourth year, I felt my confidence falter. I had just received a series of disappointing messages from the authorities holding our land use permits. They were under pressure to cancel my certifications, and if that happened, the wolves would have to go. We were still searching for a permanent home for the pack, and the consequences of an eviction were unthinkable. Even more concerning were letters that had begun to appear in our mailbox—anonymous threats to "get rid of those wolves or we will." Someone posted signs near our camp warning us to be gone or "wind up in the Custer County jail." Even a former governor weighed in, saying our project was nothing more than "wolf propaganda" and should not be allowed to exist. Jamie and I were painfully aware that we were responsible for the lives of these wolves, but our ability to keep them safe seemed increasingly beyond our control.

With all these concerns swirling around in my mind, I decided to go for a walk to sort things out. I was hoping

Kamots would join me. He was always the most inquisitive member of the pack, so I suspected he would follow if I walked alone into his domain. Normally I would have set out carrying a camera and tripod but this time my mission wasn't to film. It was just to think.

Sure enough, as I hiked uphill toward Williams Peak, Kamots trotted along after me, although he followed at a dignified distance, pretending to be more interested in chipmunks than what I was doing. As I walked, he would disappear into the trees then reappear in front of me, only to disappear again. I got to a ridge overlooking the meadow where the pack often played. I could see the white canvas of our tents and a thin column of smoke rising from the woodstove inside our yurt. I sat down, and Kamots emerged from the forest. He solemnly walked over and sat down, facing me.

We had accomplished so much together—Kamots, his packmates, Jamie, and me. We were documenting the intimate family-oriented side of wolves that people had never seen before. A new audience was developing a more accurate understanding of these misunderstood creatures. We had to find a way to continue our work and, more importantly, protect the pack that had already given so much of itself.

As I sat and wondered what to do, Kamots seemed to be wondering about me. I had the habit of talking to myself when I was trying to sort out problems, and with each sound I made, he cocked his head, the light fur above his eyes knitting together in an expression of curiosity and concern.

Then, in a gesture I'd never seen before, he raised his paw up to me. I put out my hand and pressed it against his paw, and we sat there like that for a minute in silence. I felt as though he was assuring me that if we held up our end, he would hold up his. We should continue to deal with the human world, and in his calm, strong, confident way, he would keep his pack stable and safe. He was that kind of leader.

CHAPTER FOUR

ONE FOR ALL

Jamie

On a dismal February morning our taxi came to a stop in front of what we believed to be our destination. Peering from the backseat through the pouring rain, we could only see a massive brick multigabled building, shrouded by leafless trees.

"Jim, is that it?" I laughed. "I think we're at Hogwarts!"

I am the product of an all-American suburban public school, and with its collegiate gothic facade, the Taft preparatory school in western Connecticut looked like something from another world to me—stately and yet somewhat menacing. Alas, instead of Albus Dumbledore, a smiling headmaster escorted us to Bingham Auditorium, where row after row of typical kids were quietly shuffling in to fill the seats.

We had come to Taft to deliver our multimedia presentation for the entire school. It's what we've been doing in the years since we said goodbye to the Sawtooth Pack. Sharing with others what they taught us is our way of keeping their

An exuberant pack rally

memory alive and honoring our commitment to the wolves who still struggle for survival today. The students had been required to read *Never Cry Wolf* by Farley Mowat over their summer vacation.

Many of them had seen our films, and a few even welled up with emotion as we greeted them before they filed into the auditorium.

The Sawtooth Pack has touched a lot of hearts, but even in the 21st century some children know more about the "Big Bad Wolf" of fables than they do about the real thing. That is why we speak at schools. Farley Mowat's wonderful writing may have been their first exposure to the curious, gentle, family-oriented animals who Jim and I have come to know. We were eager to introduce them to Kamots, Lakota, Chemukh, and the rest of the pack through our films, photographs, and stories, and to talk with the students about comparisons between the Sawtooth Pack and Mowat's wolves of subarctic Canada.

The children of the Taft School were well prepared for our visit and took their seats quickly. Then as we stepped onto the stage, a familiar yet decidedly out-of-place sound began to rise from the auditorium. More than 600 students were howling in unison! We stood there spellbound, enjoying how much they were clearly loving the outburst. Eventually the headmaster decided the welcome had gone on for a respectable length of time and tried to quiet the students, but a pack howl is a contagious event. It took a while to calm everyone down.

I came to learn from the teachers that this wasn't an orchestrated event but a spontaneous gesture of welcome that had spread through the room. They were excited about our visit, and their howl seemed to channel that excitement. What was especially wonderful to see was how much the kids genuinely loved howling together. The fact is: We're both social animals—people and wolves. This kind of group behavior reinforces a sense of unity, cooperation, and togetherness, and it just feels good.

I often like to say that wolves howl for more reasons than we will ever know. The Sawtooth Pack sometimes howled in celebration after a meal. They sometimes howled back and forth at night to communicate to one another, as if to say, "I'm over here; I'm okay." They howled mournfully in sadness when one of their family members died. Sometimes it seemed as if they howled just because they felt it was time to howl. Although their howls carried many different moods, the one quality that always seemed to be present was a sense of togetherness. I'm certain that wolves enjoy that sense of togetherness as much as those Taft students did.

Through our years living with the Sawtooth Pack, we often saw the wolves behaving with what seemed to be a common mind. One of the most astonishing scenes Jim and I ever witnessed involved the watchful wolf Motomo, this time in partnership with his brother Amani. These two had an interesting relationship that we could never entirely work out. Amani was sometimes pugnacious and often tried to make a show of dominating his brother. Motomo, on the other hand, was cool and never seemed very perturbed by these displays—except when there was food around. When a deer or elk carcass appeared, Motomo refused to back down, leading us to believe that despite Amani's showmanship, Motomo was probably the more dominant of the two, but he only bothered to show it when it mattered.

On one particular February day these two put aside their differences and appeared to hatch and execute a plan to get food away from the alpha. That morning we had brought the pack a deer that had been killed on the highway down in the valley. As always, Kamots took it upon himself to look after the youngest members of the pack, making sure that the yearlings had enough to eat. Later that day we had decided to give them another very small deer. We had thought we would save it for another time, but then decided it was too small to be a meal in itself. With all the chaos of snarling and growling around the food, Lakota and Chemukh were too timid to try to grab a bite. Motomo and Amani were torn between fearfulness and eagerness. They sniffed around the scene, snatching up any scraps that had been inadvertently cast aside, but they were not quite able to get in and find a spot on the deer. Evidently they were still hungry, and this bonus meal got them in a competitive mood.

The small deer was disappearing fast. All that remained were two disembodied hind legs and a portion of the torso, and Kamots and the youngsters appeared to be intent on finishing it all. As I was recording them, I could hear Motomo and Amani whining back and forth to each other in what sounded like frustration. Perhaps their conversation was deeper than that, though, because in unison they turned toward Kamots as he ate. As they faced him, he uttered his warning growl, telling them to not even think of approaching the deer. They continued to whine to each other as they watched him.

The scenario that followed occurred with such incredible speed that it stunned Jim and me as much as it did Kamots. Looking as if he had been shot from a cannon, Motomo leaped directly at the carcass and grabbed a small chunk of hide and meat. It was barely worth the trouble, but Kamots took the bait and chased after him, momentarily leaving the carcass unattended. Seizing his moment, Amani rushed in and snatched one deer leg and made a beeline for the willows.

Kamots was distracted by this new attack and turned to chase after Amani. Now it was Motomo's turn again. He dropped his small morsel of meat (which Lakota eagerly gobbled up), swung back around toward the carcass, grabbed the other hind leg, and veered off in a direction opposite to where Amani had gone. This final maneuver bewildered Kamots so completely that he broke off his pursuit of Amani and didn't even bother to chase Motomo. He walked back silently to stand guard over what had become a nonexistent meal.

Jim and I couldn't stop talking about this interaction for days afterward. It all looked so perfectly coordinated that random opportunism seemed out of the question. The way they approached Kamots together, the unusual vocalizations they made to one another, the way Motomo dropped his "decoy" piece of meat the instant Kamots's attention turned toward Amani—all these factors indicated to us that Motomo and Amani had hatched a scheme and deftly pulled it off.

Had they been hunting for an elk or moose, Motomo and Amani would have probably been the grunts—the mid-ranking foot soldiers. They wouldn't be in charge, but they had the experience to get the job done. Part of that experience would be a knack for reading each other's signals and taking each other's cues. They would be working together as a coordinated team, just as we watched them do.

We often hear the term "lone wolf" in reference to a person who acts alone, cares for no one, and craves no companionship. Sadly, like so many negative things falsely connected to wolves, the term has now come to mean a loner who wishes to do us harm. Such a state is an aberration among humans, and it's equally rare among wolves. In nature, a lone wolf is a temporary phenomenon—what biologists call a disperser. Most often it's a wolf in its third year or so who has decided to leave its birth pack and strike out in search of new territory and a mate. What does a lone wolf want? It wants to stop being a lone wolf. It wants togetherness, to be a part of something bigger. Survival depends on it.

In a pinch a lone wolf can survive by hunting field mice, gophers, and ground squirrels, but lone wolves have rarely been known to tackle large prey on their own. A solitary mountain lion has the power to crush the windpipe or break the neck of a 700-pound bull elk with its jaws and claws. One swipe from a grizzly bear's powerful paws could take down an adult moose. Wolves don't have especially powerful muscles or sharp claws. But they do have a few advantages—persistence, intelligence—and each other.

Wolves survive by chasing down large prey as part of a team. Observers of wolves have reported that although most wolves in the pack are present for the hunt, fewer than half actually take part in bringing down the prey. The youngest wolves frequently do nothing more than observe and learn from the sidelines. Each of the other pack members contributes according to its particular experience and ability. Once the pack has selected a target, the speedy, lightly built females and young males often do the bulk of the chasing. They dart back and forth, trying to grab a hind leg, slowing the prey down and giving the other wolves a chance to move in. Big alpha males frequently lag far behind during the early phase of the chase, but then when they do catch up, they often attack the prey head-on and deliver a takedown blow.

Equipped only with legs for running and jaws for biting, wolves need to make the most of their limited assets, working as a team, searching for the weak or injured animal in the herd. They don't rely on ambush or physical strength, so it's not uncommon for a wolf to be seriously injured by flailing hooves and slashing antlers. Wolves are very good at picking up on what another wolf is doing, so it's not surprising that they would be good at cooperating. I've spent years recording the vocalizations of the Sawtooth Pack, but I have also observed that most of their communication involves posture and other visual cues. Anyone who has dog companions understands that they are very good at reading body language, posture, and movements. A dog is able to tune into so-called "gaze cues," essentially watching where a human looks and turning its attention in that direction. It's very likely that wolves hunting in a pack use the same cues. It seems instinctive for them: I was always amazed to see that whenever I came upon one of the Sawtooth Pack on its own, it would look straight into my eyes as if trying to read my thoughts.

We would often catch one wolf or another observing us covertly, in what seemed to be a desire to know what we were up to. The dark wolf Motomo was especially fond of watching us carry out our daily activities. Sometimes I'd be shoveling snow off the yurt platform and Jim would be chopping wood a hundred yards away. Jim and I couldn't see each other because a patch of evergreens stood in the way. Motomo would station himself off to one side, somewhere in between, where he could watch us both at the same time. It seemed important to him that the little human team of two was still together and accounted for, but more than that he just seemed to be trying to figure us out and understand our actions. We sometimes wondered who was studying whom.

Recently Hiromi Kobayashi and Shiro Kohshima of the Tokyo Institute of Technology, as reported by anthropologist Pat Shipman, put forward an interesting hypothesis suggesting that the sclera of the eye may be involved in this type of cooperation. Although we commonly call it the "white of the eye," it isn't white in many animals. Even in our closest primate relatives, a light-colored sclera is highly uncommon. But humans, wolves, and dogs have true "whites" in their eyes. In canines the sclera is not usually visible when the animal is looking forward, but it becomes visible when they look to the side. The main advantage to having white sclera, so the theory goes, is that the contrast with the darker iris of the eye makes it very easy to deduce where the other person (or the other wolf) is looking. Developing a white sclera may have helped both wolves and early humans hunt silently and cooperatively. With just a look, a hunter can easily direct the attention of his comrades toward prey or danger. What's more, the ability to pick up the gaze cues of another species may even have contributed to the hunting partnership between wolves and humans that created the domestic dog.

IT SEEMS STRANGE TO ME THAT PEOPLE should vilify wolves for the teamwork that makes them so successful. Yet even before the colonization of America, it was common to hear wolves described as "cowardly" animals that descend upon their victims in a savage mob. And still today you hear references to violent street gangs as "wolf packs," a term chosen to accentuate the horror that they inflict. For some reason we look upon the cooperative group hunting strategy of wolves as less honorable than the swift and brutal ambush of a mountain lion, but why? After all, our human ancestors lacked the speed and power to hunt alone, and we achieved an advantage in teamwork in exactly the same way wolves do. We attacked our prey in groups and likely killed it with multiple blows instead of a single deadly strike. Perhaps when we fear wolves, it is our own past we fear. Or perhaps it makes us uncomfortable to acknowledge the deep vulnerability of being human. As much as we admire individual prowess, we are only as good as our collective effort.

Of all the opportunities we have to understand wolves and ourselves, this may be the most significant. Wolves need each other just as humans need each other. We have always known this, and the wisest among us have known that this is something to be celebrated. Over a hundred years ago Rudyard Kipling acknowledged this fundamental truth in "The Law of the Jungle" from his classic collection of fables, *The Jungle Book*. He was writing about the wolves of India, but of course he was writing about us too:

> Now this is the law of the jungle, as old and as
> true as the sky;
> And the Wolf that shall keep it may prosper, but
> the Wolf that shall break it must die.
>
> As the creeper that girdles the tree-trunk, the
> Law runneth forward and back;
> For the strength of the Pack is the Wolf, and the
> strength of the Wolf is the Pack.

We give what we can of ourselves to our family, our friends, our team, and our tribe. The benefits we receive in return are doubled—in strength, security, and survival. When a group howls together—whether they are a pack of wolves or a group of school students—they are celebrating that timeless and wonderful truth.

CHAPTER FIVE

NEVER STOP PLAYING

JIM

B Y MID-NOVEMBER, THE BRIGHT ORANGE AND GOLD of autumn had long vanished from the Sawtooth Valley. Mountain chickadees flickered among the bare red willow branches, looking for the last of the insects. A thin film of ice had begun to form along the edges of the creeks and on the surface of the riparian pond next to our camp. The past several days had been uncannily still, as though all of nature was waiting for the inevitable arrival of winter. The wolves had been preparing for it too. Close to their skin, a thick, woolly undercoat had developed to keep them insulated and warm. Longer guard hairs on top of that would keep the wind and snow at bay. Their heavier coats made them look much larger than they did in summer, and their individual markings became much more distinctive and dramatic. I always preferred to film them in cold weather, when they looked their most "wolfish."

Siblings Wahots and Wyakin

Clouds began to swallow up the peaks of the Sawtooth Mountains, intensifying the feeling of stillness. At last, huge, wet flakes tumbled out of the sky. In a matter of minutes snow was coating the straw-colored grass and boughs of the spruce trees. From the mountains to the valley floor, everything turned white.

From the looks of it, the wolves of the Sawtooth Pack had been eagerly anticipating this moment for months. First I saw Lakota dashing out of the forest, with Motaki in hot pursuit, nipping at his heels. Then came Kamots, bounding into the clearing, snapping at the falling flakes and rolling ecstatically in the first snow of the long Idaho winter.

Lakota turned to face his brother, Kamots. With a bounce, he bowed his head to the ground, forelegs splayed wide and hind end pointed high. It was the classic play bow, an open invitation to have some fun. Kamots was already gripped by the spirit and didn't need to be asked. He lunged at Lakota and the two took off in a tear, sprinting a full circle around the clearing before returning to roll in the snow with their packmates.

As a filmmaker, I was also having a wonderful time. The wolves were ignoring my presence and were completely focused on each other and the moment—or so I thought. I was intently trying to film their behavior while keeping my camera dry with a raincoat, peering through my lens and carefully adjusting the focus on Lakota and Motaki as they raced through the falling snow.

I failed to notice Kamots, creeping in from the side. I was totally unaware of his presence until I felt the first tug, and by then it was too late. Instinctively I grabbed the tripod to keep the camera from toppling over, but I was unable to save the raincoat. Immediately it became the object of a game as each wolf tried to snatch it away from Kamots. Lakota closed in and grabbed hold of one of the sleeves. The jacket seams held tough for a surprisingly long time—a testament to quality construction, I thought. But these are animals who can dismember an elk carcass in a matter of minutes.

It was a twofold loss. Not only did Kamots steal a very nice rain jacket, but he also ruined a wonderful filming opportunity. Of course he didn't know what had made the scene so extraordinary to a filmmaker's eye, with the diffused light so beautiful and everything looking like a fairyland with enormous flakes swirling about. Meanwhile, all I knew at that moment was that the wolves were tearing a man-made object to shreds—not at all the wild behavior I wanted in a wildlife film. But they were certainly having a great time.

The wolves were youngsters that autumn—before we added Matsi, Amani, and Motomo to the pack—but even as they grew older, they never failed to greet a snowfall with pure joy. Nor did they lose the mischievous streak that cost me more than one personal item over the course of the six-year project. No matter how old wolves get, they never stop playing.

Of course, no time in a wolf's life is more devoted to play than puppyhood, and at no other time is play more critical. We raised all the Sawtooth pups by hand and, like a mother wolf using her tongue to clean them, we would stroke them with warm, damp rags after they ate. Often we found ourselves in the middle of a tug-of-war over the rags, especially with Wahots and Wyakin, the brother-and-sister duo from the third litter. Those two were always full of mischief, and loved to take our attempts to clean them up as a chance to play. With all the ferocity of adult wolves, these two pups would bite on to a wash towel, growling and shaking it as if they were bringing down a bull moose. They were so adorably feisty that we were tempted to be drawn into their playtime and roughhouse with them as if they were dog puppies, but that would have been against the standards I had established at the beginning. We wanted these wolf pups to accept us as a trusted presence in their lives, but we didn't want them to look upon us as playmates.

Very young wolf pups face many risks that larger adults can ignore. Chief among these are great horned owls, which could snatch a wolf pup in the night. For their safety, Wahots, Wyakin, and Chemukh, like the older wolves before them, spent their earliest days at our home in Ketchum, an hour's drive south of wolf camp. By mid-June we began taking them with us on trips to camp to introduce them slowly to their new extended family. We configured a small pen to fit in the back of our van so the pups could stay together for the trip north. Domestic dog puppies are known for their innate enthusiasm for car rides, but I had learned with the previous litters that five-week-old wolf pups were not the same. While I drove, Jamie had the job of sitting in the back with the pups. Wahots, Wyakin, and Chemukh were the first litter that we had raised together, and I warned her, based on my earlier experiences, that the three pups might be affected by the motion of the van—but I'm not sure Jamie took me seriously.

We were barely out of the driveway when Jamie gasped. Wyakin started to vomit, and then Chemukh and Wahots, one by one, like exploding bottles in a bootleg brewery. After 20 minutes of driving, the road began twisting its way over Galena Summit in a series of hairpin turns. We had barely begun to climb when I heard Jamie exclaim again from the back. I knew from experience what was happening. No sooner would one pup finish than the next one would begin, and when they all had taken a turn, the first one started again.

I should mention that we had been feeding them a mixture of raw chicken and special puppy formula. The mixture was enough to make one gag the first time around. Reappearing in a homogenized state and flowing across the newspaper-lined floor of the pen, it was a bit more than Jamie could take. From the front of the van, I could hear the pups snarling playfully as they tore up the newspaper and spread the mess around. Jamie uttered a few phrases not appropriate for this book. Whenever she reached into their pen with a handful of fresh newspaper, Wahots, Wyakin, and Chemukh grabbed it and tore it to shreds, just having a marvelous time. When we arrived at camp, the pups were perfectly happy and ready to play some more. Jamie walked wordlessly to our tent and disappeared for an hour.

WAHOTS AND WYAKIN WERE INSEPARABLE PLAYMATES, and as they got older they developed a little comedy routine that they'd perform every time we fed them. Wolves are gorgers. When they get food, they'll eat as much as they can, because they can't be sure when the next meal is coming. It's especially visible when pups eat. Their bellies inflate like basketballs, and they stumble around in a daze of contentment. So toward the end of every meal, when the food was nearly gone, Wyakin would be blown up like a tick, but she was not content simply to fill her belly to near bursting. She began collecting chunks of meat and carrying them off into the willows to hide them. In her greed, she kept trying to carry more than she could possibly hold in her jaws. Halfway to her hiding place she would drop some, and when she'd try to recover it, she'd inevitably drop another. I used to call her the "laundry lady" for the way she waddled toward the willows, dropping things as she went.

Wahots, as always, watched his sister intently. He had learned about his sister's tricks early in life, and he knew that if he sat back and waited, Wyakin would amass a bountiful supply of leftovers for him. He would shift his gaze between his sister and the rest of the pack, lest one of the others catch on and spoil his fun. Every time Wyakin returned to the carcass for another load, Wahots would step away with an air of nonchalance, only to return a few minutes later looking as though he'd just been for a casual stroll. Then Wyakin would spend the next hour scouring the willows, trying to figure out where her food went. The siblings played out this scenario time and time again, and Wyakin never did catch on—though there were times I could have sworn she must have known and that this was all just a big game between the two of them.

Wahots and Wyakin were especially fond of jaw sparring, in which one wolf tries to clamp down on the snout of the other. It's fun for them, but it's also a subtle way of asserting dominance. In this case, play has a serious side: By the time they're juveniles, pups have sorted out their social structure, mostly by playing with each other.

As young wolves are building pack bonds and working out their pup hierarchy, they're also developing the skills they'll need to survive and become contributing members of the pack. Most of the play we witnessed involved a lot of running around, and that's hardly an accident. There's an old Russian proverb that goes "The wolf is kept fed by its feet." The ability to chase after prey, sometimes for several miles, is the most critical survival skill a wolf possesses. Wolves, especially young ones, appear to run just for the pure joy of it. The Sawtooth Pack could spend hours chasing each other back and forth through the trees in a lively game of tag. There's an evolutionary advantage to this childhood behavior— they are building speed and stamina—but it always just seemed like they were having fun.

If tag was their favorite game, keep-away was a close second. One wolf would pick up some random item and begin to prance about, flipping it up in the air, daring the others to try to take it away. This would usually end in another chase or tug-of-war for possession of the fetish. Just about anything could serve as a toy—a bone, a stick, even a pinecone—but nothing matched our human possessions. We tried to be vigilant and keep all our gear carefully stowed away, but wolves are very observant and people are easily distracted.

I was amazed at the things they managed to steal in a split second. Once they managed to grab a sledgehammer and had a marvelous time dragging it around the meadow, losing their grip, tripping over it, and stealing it from one another. Another time Matsi pulled a heavy toolbox out of a cart, spilling its contents in a heap. The crash sent him running away in surprise—and gave us the opportunity to retrieve our tools.

Their obsession with human objects sometimes caused embarrassing situations for me. In the project's third year, Ted Koch, an official from the U.S. Fish and Wildlife Service, paid a visit to our camp. Ted was recently assigned to be in charge of wolf recovery in Idaho, but wolves were so scarce in the wild that he had never actually seen one. Visiting the Sawtooth Pack gave him the opportunity to get a close look at the animals he would be managing. I was overjoyed that my project could be an asset to the wolf recovery effort, even before our films came out.

At the time of his visit, though, I was in Washington, D.C., extending our land use permit with the U.S. Forest Service, so I could not meet him at wolf camp. In my absence our assistant, Val Asher, took him into the enclosure. As the story goes, she did advise him not to put any of his gear on the ground. Ted had brought a small 35-mm camera as well as a video camcorder. When he first saw the pack, he began snapping photos, but when the wolves came closer, he put his still camera down and began videotaping. A moment later, Kamots and Matsi were frolicking a short distance away, playing keep-away with a small, black object. Ted filmed them happily for a minute, then looked up from his camcorder and innocently asked, "What is it they're doing?"

Val took a good look and then said, as casually as possible, "They're eating your camera."

Ted was so captivated with the pack that he took it in stride. Weeks later, while walking through the meadow, I found a small chunk of plastic, about the size of a half-dollar, with a few shredded wires dangling from it. I put it in an envelope with a note that read: "Dear Ted, Kamots enjoyed your camera. He's finished with it now."

AS THE WOLVES OF THE SAWTOOTH PACK GREW from pups, to adults, to older wolves, their appetite for play never diminished. In fact, I'm sure that, just like pups, adult wolves play for the sheer joy of it. But there's a lot going on in a little game of tag. For animals that live and hunt as a group, a solid foundation of trust and cooperation is critical. A wolf pack succeeds or fails as one.

Through play, adult wolves stay in tune with one another. After a good chase or a tussle over a bone, wolves have a better sense of one another's physicality, strengths, and weaknesses. It isn't that much different from a football team running practice plays. The more they move together, the better executed their plays become. Plus wolves have a great time doing it, which builds camaraderie and reinforces their pack bonds.

There's another side to play that's harder to define. It's something Jamie and I came to understand as we observed the Sawtooth Pack. A wolf pack is indeed a family, but it's also a hierarchy. We saw the wolves use play as an opportunity to put that hierarchy aside, if only for a short while. Lakota, the pack omega, was most often the one who instigated play. As the omega, he often had to endure dominance displays from the rest of the pack. But he was also a bit like a court jester. It stands to reason that the lowest wolf in the hierarchy would want to keep things light and relaxed. And so if social relations got tense, Lakota would be the first one to bow his head, wag his tail, and say, "Let's play!"

I remember a moment when the wolves weren't doing much of anything interesting and I was simply watching them, as I often did, hoping for a bit of behavior that would make a good filming opportunity. All of a sudden Lakota walked over to his brother, Kamots, and crouched into the characteristic play bow. It was as though Kamots had been waiting for a trigger. He sprang to his feet and darted after Lakota, snapping playfully. Soon they were coursing through the meadow, Lakota only inches ahead of Kamots's jaws. Lakota had had his share of other wolves snapping at him, but this time it was clear from his posture that he understood this was not aggression.

Lakota zigzagged through the grass with his mouth agape, his lips pulled back in a wolf smile—that unmistakable expression of joy that we humans seem to understand instinctively. Several times he nearly allowed himself to be caught, only to leap just a hair ahead of Kamots. Eventually Kamots did catch him—or maybe

Lakota let himself be caught. He flipped over onto his back in surrender, while the alpha snarled and snapped in mock ferocity. Then Lakota gently licked Kamots's muzzle—the wolf equivalent of saying uncle—and the game was over.

In this case Lakota and Kamots played their roles in accordance with pack hierarchy, but after observing them do this over and over, I began to see that it wasn't so simple. Not long after I had observed this interaction, I again watched the pair racing through the meadow. The two brothers are extremely similar in appearance, large with classic black and gray markings, so it took me a minute before I realized that this time Lakota was chasing Kamots. I couldn't imagine how this role reversal could be happening. In an amazing twist in their standard game, Kamots was allowing himself to be caught in the game of tag.

Yellowstone's Rick McIntyre, the most seasoned wolf-watcher in the park, shared a similar observation with us. He had once watched 21, the fabled leader of the Druid Peak Pack, mock-lose a fight with his pups, even rolling over in apparent submission. Although he was a big, powerful wolf and the top of the Druid hierarchy, he was willing to play a game to help build his pups' confidence and instill a cooperative spirit within their pack.

On the surface this kind of role reversal might seem like a simple thing, but it implies that there is a lot more going on in the inner life of wolves than we can imagine. In humans, an older, stronger brother might let a younger brother pin him in a mock wrestling match, feigning defeat and letting his younger sibling celebrate victory. It's the same with wolves. Both participants know who the dominant one is, but it can also be fun for both to reverse the roles. Such behavior speaks volumes about both the self-awareness and the compassion of a wolf like Kamots. It means he had to have a concept of himself and a concept of how Lakota perceived him. He must have understood that it would be rewarding for Lakota to enjoy a moment of victory, even if it was all a game. It was important to Kamots that Lakota knew he belonged to the pack as much as any mid-ranking wolf. Watching the two of them play like this, I understood how much these wolves really cared for each other.

Knowing that a wolf has a self-concept and a concept of the other wolves in the pack is fascinating enough, but beyond that, I wonder if wolves understand their place in the larger world too. Wolves don't limit their choice of playmates to other wolves. The Sawtooth Pack shared their territory with us, with various rodents, with songbirds, and with one other constant avian companion: the ravens. Biologists have written a great deal about the relationship between wolves and these extremely clever birds. Ravens know to follow wolves on the hunt because they will likely be rewarded with scraps after the wolves finish eating. It has been reported that ravens return the favor—guiding opportunistic wolves to animal carcasses and thus an easy meal. Because ravens have a bird's-eye advantage for spotting prey on the ground, wolves have learned to look to ravens to guide them to a meal. By way of reciprocating, wolves tear through tough hides that the birds cannot penetrate. Both reap the reward of a team effort. However, the two species seem to share a camaraderie that goes beyond mutual benefit. We often saw ravens hanging out near the wolves even when no food was to be had. Both species are highly intelligent, social, and communicative—really, they have quite a bit in common.

And yet the ravens did not share any of this spirit of camaraderie with us. Wolves had passed the test eons ago, but people, it seemed, were not worthy of the ravens' trust. They were skittish and never let Jamie or me get close to them, but at least we were allowed to observe at a distance the bizarre relationship that developed between the Sawtooth Pack and the resident ravens. Sometimes when we were feeding the pack, the ravens would gather nearby and keep a close eye out for any stray morsels. Every so often a bold individual would dart in, and one of the wolves would turn and snap furiously at the bird, perhaps removing a tail feather but never inflicting any real damage. I think it was mostly for show. The wolves' attitude seemed to be, "We're predators; we have to try to catch you, and you have to fly away." The ravens at wolf camp actually seemed to enjoy flirting with disaster, making a game out of seeing just how close they could get to a wolf and still escape. I believe the wolves knew it was a game too and enjoyed it just as much as the birds did.

One day Amani was snoozing in the grass when a raven landed a few feet away and began bobbing and strutting back and forth. There was no food in sight; this was purely a social call. At first Amani showed little interest, so the raven began taunting him with caws and cackles, even pulling the sleeping wolf's tail. Amani's ears perked up and he slowly shifted his weight, trying to maintain the illusion of relaxation while subtly positioning himself to move. Suddenly he pounced and the bird leaped backward, flapping wildly. Amani twisted his neck this way and that, snapping and missing. The raven sailed up about 10 feet and then glided back to earth a short distance away. No sooner had he landed than he began to cackle and hop about. Again Amani feigned disinterest, and again the bird came closer. Once more the wolf pounced, just a feather's width off the mark. This game went on for almost 20 minutes before Amani finally stood up, bored, gave a shake, and walked away, while the bird cawed after him.

On one level this simply could have been Amani's predatory instinct at work, but I don't believe he really wanted to catch that raven. In fact, although we saw plenty of interactions between the pack and ravens, we never saw a wolf catch one in six years. On two occasions we did find a dead raven lying in the grass, though. But what's interesting is that the wolves never consumed the dead ravens,

nor did they treat the carcass as a toy the way they sometimes did when they managed to catch a mouse or a vole.
Once Jamie picked up a dead raven by the wing and tossed
it on the ground in front of Matsi to see what he would do.
He sort of looked at her as if to say, "How sad," and walked
away. If a wolf had killed that bird, I can only imagine it
had been an unfortunate accident.

SEEING JUST HOW MUCH the Sawtooth Pack played, we
wondered about their wild cousins. Were wolves in
the wild, who were more burdened by survival issues
such as being hunted and avoiding rival packs, still
likely to play? Our friend Gordon Haber provided an
answer. After spending a lifetime observing wolves in
Alaska, he observed in his book, *Among Wolves,* "If a
half hour passes without at least some play, it is an
unusual half hour in the daily routine of a wolf family ...
It isn't coincidental that wolves are at the same time
probably the most playful, as well as the most socially
cooperative, of nonhuman animals." Jamie and I both feel
that Haber was ahead of his time in how he understood
the social order of a wolf pack. Where other biologists
saw a family group of hunters, Gordon saw a shared
culture and the passing of information over generations.
He saw a true society in which play is the glue that holds
it together.

Jamie and I were thrilled to have the opportunity to join Gordon on an expedition to film wolves in Alaska in 1995. The first surprise was the Alaskan weather. We arrived in a place called Delta Junction about a hundred miles southeast of Fairbanks on an unseasonably warm March day. We decided to go for a 30-minute run, wearing our light jackets, and before we made it back to our lodging, the temperature had dropped from about 40° above zero Fahrenheit to 40° below!

We were scheduled to fly in a small airplane over the Yukon River and look for wolves from the air, but when our pilot saw the gear we had brought, we had to delay and get properly equipped. Jamie and I had brought all the layers of clothing that got us through winter in the Sawtooths, so why should we worry about being cold in Alaska?

Our pilot had a treat for us filmmakers. When we got to the plane, we saw it had no door. That way I would be able to film wolves without having to peer through scratched Plexiglas windows. I'd have a great view, but the temperature inside the plane would be as frigid as the summit of Denali in a blizzard. That realization sent us to the nearest store in search of as many hand and toe warmers as we could find, plus a stop at the pilot's home to borrow more gear—the same gear that pipeline workers use to survive prolonged exposure to the fierce cold of the tundra. By the time we boarded the plane, we were buried inside huge parkas, fur-lined pants, and enormous "bunny boots" that made us look like we had Mickey Mouse feet. We were prepared; the only problem was we could barely move.

This was a decade before sophisticated aerial mounts were widely used, and the miniature camera drones that everyone seems to have nowadays were years from being invented. I was tethered to the aircraft with a safety belt, holding a heavy 16-mm film camera, leaning out the open airplane door and taking the full blast of 100-mile-an-hour Arctic air in the face. But what a view!

Miles from anything, high over Alaska's remote backcountry, we spotted a pack of about 15 wolves on a bald mountain slope. As they came into view, we could see them racing about, darting back and forth, moving up the mountain slope, then back down. They weren't hunting and they weren't patrolling their territory; they were playing. As we got closer, we could pick out one wolf in particular that seemed to be the master of ceremonies. He would run up the side of the slope, then run back down at breakneck speed, chasing one or another of his packmates. Then he'd tumble and roll in the snow, leap to his feet, and drop his head into a play bow. He seemed to be able to get every other wolf into the mood. They all took turns bounding after him through the deep drifts. Although it was too far to see, I could imagine their wolf smiles as they chased each other and nipped at one another's tails.

Watching these wolves, we thought right away of our Lakota. We knew that this exuberant ringleader below us was the omega, the one who got the game going and invited the others to play. He might get picked on occasionally, but he kept the mood light and helped all the wolves stay close and bonded to one other—truly an important and cherished member of the pack whose importance came in large part because he was always ready to play.

CHAPTER SIX

TEACH THE YOUNG, RESPECT THE OLD

JAMIE

WAHOTS, WYAKIN, AND CHEMUKH, four-month-old wolf pups, were heading off on an adventure. It was a sunny morning in August 1994, just one day after Jim and I had introduced these three into the pack. Matsi, the pack's beta male, had stepped up to be the primary sitter and caregiver to his newly adopted packmates. Almost from the moment they met him, standing calmly, ready to greet them outside our camp gate, these awkward little pups seemed to understand that Matsi was their protector and mentor. They followed him immediately. They hadn't chosen Matsi; he chose himself. He had this

Matsi puppy-sitting Wyakin and Wahots with Chemukh behind

calm, even-keeled presence that must have been reassuring. Even shy Chemukh appeared to feel safe and comfortable around him. Although Kamots was the first to welcome
the new pups into the pack, it was Matsi who took them
under his wing.

From our small camp in the middle of their territory, we
could witness this fascinating behavior without interrupt-
ing them. We were unobtrusive observers living in a little
bubble within their kingdom. This idea of ours—to put our
camp inside their territory—was paying off more than we
ever expected. Our presence became a normal part of their
day-to-day lives, and we were rewarded with unparalleled
access to every aspect of their society.

Before Jim began this project, many people thought that
a wolf pack was just a group of individuals who stuck
together for mutual benefit. When the Sawtooth Pack came
into his life, he realized that they were so much more. A
pack was an extended family, wholly devoted to one another,
bound together by a common purpose and at times what
seemed like a common mind. Before he understood this,
he worried that one of the wolves would try to climb the
fence and run away. A bold wolf, he thought, would find
the call of the wild impossible to resist. Ultimately, we found
the exact opposite was true. Although certain individual
wolves can be gripped with the urge to wander, no member
of the Sawtooth Pack ever tried to leave the enclosure, nor
did we ever see them restlessly pace the perimeter. Their
spacious 25-acre territory was an ideal place to live, but we

don't think that was the reason for their contentment. The reason was each other. Each wolf knew that it belonged with the pack, with the family.

On his second day as pup-sitter, Matsi must have decided that his duties should include showing the pups around their new territory. He began by leading them out of the meadow near our camp to a marshy forest of willows and then up into a dark woodland of pine and aspen, where they briefly disappeared. Their two-year-old uncle, Amani, also tagged along, apparently curious about what they were up to. The rest of the pack, exhausted from yesterday's meet and greet, continued snoozing in the warm summer sun on the edge of the clearing.

At nearly every other step the pups wanted to pause and sniff the wet ground or chew on willow twigs or pinecones. Matsi waited patiently for a moment or two as they did this. Then he'd continue walking, and they'd bound after him. He led them in an arc around our camp, eventually ending up at a small pond near the north end of their territory. It was a pond that Jim had renovated, back when he was just starting the wolf project. He had found the remnants of an old beaver dam long abandoned by its builders. It only took putting some of the beaver-cut sticks and a few rocks back into place for the pond to refill. The result was a lovely water feature and one of the wolves' favorite places to play and explore, especially in the warm days of summer.

The pups hadn't seen anything like a pond or a flowing creek before, and they were at once fascinated and a bit tentative. They stopped frolicking and huddled close to Matsi as he walked to the edge and waited patiently for them to investigate at their own pace. One by one they summoned the courage to dip a paw, then a nose in the cold mountain water. Immediately fear turned to delight as they began slapping the surface with their forepaws, then darting away as though the water might chase them. The excitement of the new discovery prompted Wyakin to pounce on her brother, Wahots, and the two rolled around on the muddy bank in a joyful snarl. Chemukh, always the timid one, held back from the roughhousing, but she bounced and splashed by herself along the water's edge.

All the while Matsi stood stiff-legged and attentive, not participating in playtime but seemingly content to watch the pups become comfortable with the new environment. After 10 minutes or so of watching them retrieve floating sticks, Matsi turned and stepped out onto the rickety beaver dam and walked to the other side. The whole thing was no more than six feet across, and Matsi was on firm ground again with just a few steps. Nevertheless, the three pups promptly stopped their play and stared at him as if he had just levitated. Their little faces seemed to say, "Whoa, you can DO that?"

They had just conquered their fear of touching the water, but crossing it was a brand-new challenge. Wyakin made a tentative attempt to step onto the beaver dam, but when one of the sticks shifted under her weight, she retreated. She made little feinting steps toward the dam, but it was clear she had lost her nerve. Matsi understood her apprehension and made his way back. He stood on the dam, leaned over, and touched his nose to Wyakin's. She began to lick his muzzle in submission, as pups do. Then Matsi carefully turned 180 degrees and walked back to the other side.

This time Wyakin stepped up behind him and tentatively began to pick her way across. To be fair, it was a bigger challenge for her little legs. Every so often she lost her footing, and one of her paws would slip into a gap between logs. She'd scramble back to her feet, acting like the dam had just tried to swallow her up. Eventually she made it across, leaping over the last six inches in a break for solid ground, then tumbling into Matsi. As soon as she had made it, she became all puffed up with confidence. She turned and paced the bank, whining to her littermates as if to encourage Wahots and Chemukh to follow her. At a similar pace and with similar grace, the two other pups summoned their courage and made the passage. Their triumph sparked another round of play, and this time even Chemukh joined in.

It was fascinating to watch Wahots, Wyakin, and Chemukh explore and grow with each passing day. Watching, following, and imitating the older wolves was just another form of play to them. If an adult sniffed or touched anything, one of the pups would be there to investigate it, too. If an older wolf stepped up onto the trunk of a fallen tree and had a look around, one of the pups would have to scramble up and stand in the same spot.

We noticed that the three-year-olds, Kamots and Lakota, were always gentle with the pups, but they didn't indulge the little ones as much as two-year-old Matsi, Amani, and even Motomo did. The younger adults tolerated all manner of ear chewing and tail pulling, annoyances that their elders probably wouldn't have stood for. In fact, the two-year-olds seemed to enjoy all the attention. I've since heard from wolf biologists that it's common for young adults to regress into a bit of a "second puppyhood" when new pups appear on the scene. While the oldest wolves look over the well-being of the entire pack, the junior members act as playmates and teachers, showing pups the ways of the world and helping them to build confidence.

It's not easy to document these moments of teaching because the behavior is often very subtle. Gordon Haber recorded one of the best examples of this from a plane over a river channel in Denali National Park and Preserve. It involved a small traveling party of six wolves from the large Toklat Pack: two older females, a 15-month-old yearling female, and three pups born four months earlier. The pups were just a couple of weeks older than Wahots, Wyakin, and Chemukh were when Matsi took them under his wing. When pups reach this age, the adults move them from the den site to an area that is more conveniently located so the pack can go on hunting expeditions but the pups can stay behind in a safe place with a pup-sitter to wait for the older wolves to return with food. Biologists call this location a rendezvous site, and, just like a den, a wolf pack may use the same site year after year. Gordon recognized that because these wolves, including the pups, were on their way to the rendezvous site, this would have been the pups' first foray into the wide world beyond the immediate vicinity of their den—and their first encounter with fast-moving water.

The two older females, one of whom was the mother of the three pups, waded across the river with little apparent concern. The yearling female crossed with them but remained very attentive to the little ones tagging along behind. As the three pups hesitated at the river's edge, she turned and waded about a third of the way back into the channel, crouching into a bow and slapping the water playfully with her paw, making a game out of the frightening moment. One little black pup made it out to her but lost its nerve and retreated to the bank. Next, a little tan pup gave it a go, but it crossed too far downstream, where the opposite bank was undercut and steep. It succeeded in making it to the other side but it couldn't climb out. The yearling female saw its plight, leaped up the bank, dashed back, and grabbed the pup in her jaws and lifted it out.

By this time the black pup was trying again, even farther downstream. It was now struggling in deeper water and being carried away. The yearling wolf ran to it and jumped in just down current, blocking the little black pup from being swept away farther and letting the pup use her body as a brace to climb up and out.

All this while the third pup had been observing from the far bank. After seeing the struggles of its two littermates, it chose a crossing spot farther upstream, where the water was shallower and the bank was low. Once they had all safely crossed the river, the dominant female and her adult packmate continued the journey with little fanfare, but the three little ones gathered around the yearling, their guide, with a great deal of bouncing, and she seemed to relish their success.

NAVIGATING THEIR PHYSICAL ENVIRONMENT is only part of what wolf pups need to learn. They also need to learn how to navigate the social environment of a wolf pack. Pups begin their lives in a blissful state outside the adult hierarchy, but as they mature, the adults begin to enforce the rules.

For example, Jim and I began to notice that after Wahots, Wyakin, and Chemukh had been with the pack for a few weeks, Kamots started dominating Chemukh at mealtimes and making her wait with Lakota, the omega. Kamots appeared to single out Chemukh for extra discipline. His rule was intense. Sometimes he would leap over the kill to pin Chemukh to ground, growling and snapping his jaws as he straddled her. Chemukh would cry out and appear to yelp her surrender, and then she would retreat into the willows with her tail between her legs.

We often wondered what caused this behavior. Our best conclusion was that it had something to do with Chemukh's timidity and insecurity. If she had aggressively leaped on the kill, as Wahots and Wyakin did, perhaps Kamots would have let her eat earlier. As it was, when she approached the carcass in a nervous crouch—ears back, tail tucked, in a submissive posture—Kamots wouldn't stand for it. It almost seemed as if he was playing the part of drill sergeant, trying to toughen her up and make her more assertive. After all, it is in the best interest of a wolf pack to have all its members strong and working together as a cohesive unit. A wolf who can't work with the others is a liability. Perhaps this was his way of making sure Chemukh would grow up to be an asset to the pack.

Observing a wolf pack with pups, we became more and more aware of how obsessed the adults are with the little bundles of fur. Most of the time they are very gentle with young wolves, but watching the way Kamots dealt with Chemukh, we became convinced that the adults initiated these interactions, whether doting or disciplinary, for the sake of guiding their young on the confusing path toward adulthood.

A wolf needs to grow up and make a contribution. She may turn out to be a swift hunter; he might have a protective streak and excel at defending pack territory. An especially gentle wolf might be suited to help raise the next generation or become an instigator of play, like Lakota. A wolf can proffer benefit to the pack society in many ways, and understanding the social rules is the first step. That's not to say these future benefits play consciously on a mother wolf's mind as she plays with her pups or scolds them when they get out of line. Her actions are instinctive in wolves, as they are in us, springing from pure affection, and concern for the protection and social well-being of the little ones. All the same, this guidance, whether it comes from parents or pup-sitters, will bring tangible benefits to the pack as a whole.

As humans we instinctively understand this behavior—at least on a small scale. Sometimes it seems like we have forgotten that this simple truth applies to our entire society beyond the family. We claim to value education, but in practice we treat failure as an unfortunate but unavoidable outcome. In a wolf pack, there are no forgotten children. Every pup is worth teaching, every pup is valued, and every pup is eventually expected to make a contribution to the well-being of the pack. Whatever it was that Kamots was trying to teach Chemukh, it worked. As an adult she became a force to be reckoned with.

If young wolves represent the future of a wolf pack, then old wolves represent its past. The soul and wisdom of the pack lives in its elders. In the developed world, where we prize youth and vigor, always looking toward the next technological advance and all too eager to forget the past, the elderly are often marginalized. We tend to think of our senior citizens as a group that needs to be cared for but not necessarily venerated. How often do we acknowledge our elders as ones who remember history firsthand, as the holders of knowledge and experience, as the keepers of our culture? Wolves do, and they do so for a very good reason: survival.

Kira Cassidy is a young biologist working in Yellowstone National Park. Her research project examined the impact of old wolves in a pack, asking if they were an asset or an impediment to pack success. With 16 years of data from Yellowstone, she offered her discoveries as part of a fascinating TED talk.

Yellowstone has the highest density of wolves in North America, and because of that, conflict between neighboring wolf packs appears to be more intense than in other areas where wolves live. In that environment, certain packs command large and bountiful territories while others struggle to hold a few square miles. Kira's study revealed that a wolf pack was 2.5 times more likely to win a territorial dispute with another pack if it had at least one old wolf—that is, a wolf upwards of four or five years or older. The explanation doesn't involve superior strength or vigor: As they age, wolves may actually participate less in the physical act of hunting and fighting. Instead, Kira concludes, the reason is experience. Old wolves know the terrain and are better equipped to pick the location of engagement. Where young wolves might panic, older, experienced wolves are likely to be calmer and more able to keep their pack unified during a conflict. The pack elder may even have engaged with the rival pack in years past and know the enemy's strengths and weaknesses. In fact, smaller packs with older wolves fared better than larger packs with no elders, Kira observed. The benefits of having an old wolf actually result in a positive feedback loop: Packs with old wolves are better able to protect their pack, and a better protected pack enables more wolves to achieve old age; thus, packs like Yellowstone's Mollie's Pack are able to dominate a region for years. Elder wolves aren't pitied and cared for, they're needed.

The presence of elders is what shapes the very character of a wolf pack. Young adults bring vitality and strength in numbers, but the older wolves guide the behavior of the pack as a single cohesive unit. The longer scientists are able to carry out continuous studies of wolf packs, the more they see that different packs actually have what could be considered different cultures. Some Yellowstone packs, such as the Mollie's Pack, have developed a culture of bison hunting, using large numbers to tackle the biggest and most dangerous prey on the continent. Other packs never look twice at a bison, focusing exclusively on elk and deer.

Gordon Haber wrote extensively about the Toklat wolf pack, which had developed a unique style of hunting Dall sheep on the rocky slopes of Denali National Park. When attacked, Dall sheep instinctively run uphill and into terrain where they have an advantage. The Toklat Pack's strategy was to move up the slope unseen and then attack from above, thus cutting off the sheep's escape route. The tactic created confusion among the sheep, and the Toklat Pack exploited it time and again. Other packs occasionally hunted Dall sheep, but none did it the way the Toklats did, or with as much success. It was their culture.

In 2005, the Toklat Pack suffered a devastating blow. Knowing that a wolf pack's territory doesn't follow man-made boundaries, hunters set up trap lines just outside the borders of Denali, waiting for the wolves to step across. Eventually the Toklats did exactly that. The alpha female was caught in a leghold trap and killed, along with two younger wolves, and a little later the alpha male was shot. The six youngest wolves who remained retreated into the park.

Gordon, devastated by the loss, resumed his study of the pack, but he never again saw them hunt Dall sheep as they once did. Instead, the pack subsisted on snowshoe hare, a prey animal the youngsters were already able to hunt (and that were, fortunately, abundant that particular year). The young wolves hadn't yet learned to catch Dall sheep the way their elders had done, and so that aspect of the pack's culture vanished, erased by a few traps and bullets, perhaps never to be recovered. Gordon's story offers a clear example of how a wolf pack passes knowledge from generation to generation. When too many of its elders are killed, experience and wisdom disappear along with them.

I remember in college watching Carl Sagan's wonderful program *Cosmos* on PBS. One of the episodes was about the catastrophic destruction of the Library of Alexandria in Egypt. When the library burned, a wealth of human knowledge and learning went up in flames, and it took 500 years or more for some of it to be rediscovered. I still wonder where the human race might be if that hadn't happened. Perhaps we have to wait until the year 2500 to catch up to where we should be now. I think of the loss wolf packs have suffered in the same way. Humans in North America have killed more than a million wolves since European colonists first arrived on these shores. I wonder what those wolves knew back then—those that had lived for generations hunting eastern elk in the Appalachians or chasing bison across the Great Plains. Think of all the cultures that were lost.

BEFORE I JOINED JIM IN IDAHO, he would call me, asking advice about a particular wolf named Makuyi, the one whose eyes needed medical treatment. On one of these phone calls he told me about a curious event involving her and a male wolf, Akai, during feeding. Only many years later did I realize that this story offered another window into wolf culture and had many similarities to Gordon Haber's observations in Alaska. One of the neighboring ranchers gave Jim a steer that had died. Jim brought it in for the wolves. At the time Akai was the oldest. Kamots and his littermates were just six months old. Akai had never eaten beef at his original home in Minnesota, and he was unwilling to take a chance on the new food. The younger wolves, following his lead, would not touch it. Makuyi, on the other hand, had grown up eating beef, as well as deer and chicken. After the rest of the pack rejected the steer, Makuyi moved in and ate her fill.

I suspect that if the other wolves had been desperately hungry, they would have been more apt to take a chance on an unfamiliar meal. In general, though, wolves seem reluctant to try food that they have not eaten before. Young wolves will happily eat anything if they see an older wolf eat it, and their food preferences appear to develop out of these early experiences. The implications are certainly worth noting as we strive to find a way to live peacefully alongside wolves. I believe that if there are abundant deer and elk around, wolves will be as reluctant to kill a cow or sheep, as Makuyi's packmates were to eat that steer. As a matter of fact, we've heard accounts from ranchers who have watched wolves wandering disinterestedly among their cattle. When it came time to hunt, the wolves targeted only the deer that had been grazing among the herd.

Unfortunately, the converse situation has played out recently in the American West. The wolves who were reintroduced into Wyoming and Idaho in 1995 were mostly young wolves captured in Alberta. They had to start from nothing, building a new culture in a new land without elders to teach them. When we removed them from an established culture, we opened the door to them experimenting with unfamiliar domestic prey. In a few protected places we have permitted packs to mature and establish firm cultures, but by and large we continue to disrupt this development through hunting. It's tragic for them, but it doesn't do us any good either. Every time we kill a pack elder, we destroy the pack's guiding culture and—critical for us—its predictable behavior, and so we willfully continue to undermine the very thing that could make coexistence between our two species possible.

When wolves are allowed to maintain their pack cultures, they are so much more predictable. They hunt the prey they're used to hunting in the places they know best. They tend to den in the same locations and use the same rendezvous sites year after year. But wolves who have lost their pack knowledge are desperate and unpredictable; their behavior can be random. Without the guidance of pack elders and without the memory of their hunting culture, they are more likely to enter human territory and prey on livestock. A large, stable pack is a far better neighbor to ranchers than a fragmented bunch of young wolves.

We see this in other social animals as well. Young elephants that have lost their mothers to poaching and war become shell-shocked and aggressive young adults. Orphaned young bull elephants are frequently responsible for raiding crops, destroying property, and injuring people.

There are human parallels. In areas ravaged by war, children lose parents, friends, and homes. Their education halts, and strife becomes their teacher. They learn to navigate in an unforgiving and unstable world. Long after the conflict has ended, they remain lost and unable to integrate. Such children become easy fodder for the false security of extremist groups, maintaining a legacy of violence.

The chilling conclusion is that what we are doing to wolves, we are also doing to ourselves. When we marginalize the older generations, we lose the experience—their ledger of mistakes, successes, and lessons learned. We lose the map of the past that could help us navigate our future. If we don't look to our elders, we ignore our history and shared experience, and we end up repeating the same mistakes over and over again. If we truly cherish the young and let our elders be our teachers, we can break the cycle of ignorance and grow together.

THE SAWTOOTH PACK DIDN'T GROW OLD at wolf camp. We always knew that our film project would have to end, and that the wolves would have to move. So before the project even started, Jim had been searching for a site where he could build a permanent home for them. Staying in the Sawtooths was never an option. The Forest Service made it clear that it was under too much pressure from the anti-wolf community to continue to extend our permits. We met with representatives of the Nez Perce Tribe and reached an agreement to build a new enclosure and education center on the Nez Perce Reservation in northern Idaho. There, the

Sawtooth wolves spent their remaining years. Although we didn't get to see them every day, as we had before, we still made regular trips to renew our bonds. And every time we arrived, we would be treated to an excited greeting, as each wolf would jump up on us to lick our faces in a frenzied hello and, perhaps—or so we felt—ask the question, "Where have you been?" It broke our hearts that we couldn't tell them how much we still loved them, even though we were far apart. Years passed, but they never forgot us.

For me, it was especially heartwarming to see Lakota, the wolf with whom I had the closest bond of all, make the graceful transition into a pack elder. He had spent most of his younger years as the omega, but after the age of six or so, that appeared to change. He didn't have some great breakout moment. And I never saw him dominate another wolf to demonstrate that he had moved up in the social hierarchy. Rather, the other wolves just seemed to allow him to retire into the role, so to speak. It was as if he had lived long enough and had enough experience that the other wolves—especially the youngest ones—treated him with a type of reverence.

I took so many photographs of Lakota as a three- and four-year-old, when he was enduring life as the omega. I remember how back then his yellow eyes often betrayed the anxiety that came with his low status. When I saw him in his later years, I was happy to see that anxiousness gone from his eyes. In its place was a calm look of wisdom and, I think, a little bit of weariness.

Cared for by humans in a safe environment, Lakota lived to the old age of 12, even longer than his brother, Kamots. Younger wolves outlived him, but for me, when Lakota left this earth, I understood that the Sawtooth Pack we had loved was but a memory. He had been there from the beginning, living his life beneath the towering peaks of the Sawtooth Mountains, experiencing every moment and every season as deeply as any wolf can. He never had the responsibility of an alpha, but I believe that his kind and playful spirit held the pack together every bit as much as Kamots, the alpha, did. From his playful youth to his graceful aging years, Lakota guarded the heart and the wisdom of the Sawtooth Pack.

CHAPTER SEVEN

STAY CURIOUS

JIM

KAMOTS SAT BESIDE A FIR TREE and eyed us quizzically as we went about our work. We trudged back and forth through the snow, wheeling heavy bales of straw to a clearing in their territory. As we arranged the bales in a rectangular shape, Motomo and Amani trotted over to watch. We stacked a second, smaller rectangle of bales on top of those, and then an even smaller level on top of that. By the time we were halfway through with our job, every wolf in the Sawtooth Pack was quietly observing us. At times like these I wondered what they thought of us—always scurrying around, busy with some chore or other, never appearing to take time to play. We must have been baffling creatures to them.

Our preoccupation at this moment was the construction of an artificial hill to enhance the wolves' territory. It was early in the project, and I was still experimenting with visual ideas for the film I was creating. I was trying to solve a geographical problem. To the eye, the scene was dramatic, with a 10,000-foot mountain as a backdrop, but the peaks were too close

Motomo, watchful and curious

for me to get both the wolves and the mountain in the same shot. To do that, I needed to find a way either to make myself lower or make the wolves higher. A friend had suggested that a small hill might add an interesting perspective. Wolves love to climb up on boulders and fallen tree trunks to survey their surroundings, and they seem to enjoy getting a little bit of height when they are howling. Their territory was a gentle slope rising toward the Sawtooth Mountains. If a wolf climbed the little hill we were creating, I could film from below and cast the wolf against the alpine backdrop. I imagined how regal Kamots would look howling from the top, with the first rays of morning sun lighting Williams Peak with an orange fire. It would be a marvelous image, so why not give it a try? The worst that could happen is that it would look phony and we'd take it down.

I purchased a supply of straw bales from a local rancher, and after the first few snows had fallen, my small team and I stacked them into a pile that resembled a six-foot-high Maya temple. To give our creation a natural appearance we shoveled snow onto our structure, trying to create an irregular shape. Then, for good measure, we doused it with buckets of creek water so it would freeze overnight. The project took an entire afternoon. As the sun disappeared behind the Sawtooths and the temperature began to plummet below zero, I took a step back and admired our creation. Once we got a good snowfall, it would look like a huge snow-covered boulder. The wolves' attention had waxed and waned throughout the afternoon. But now they seemed to sense that our project was complete, and they all trotted back and stared at this giant mound that had suddenly formed in their home territory.

The next morning, after a quick breakfast, we set out to put the finishing touches on our creation. When we arrived at the clearing, we let out a collective gasp. Our hill was gone, flattened. In its place, a vast beige smear sullied the fresh snow. Evidently, after we had gone to sleep, the wolves had intensified their investigation. From the way it looked, I imagined that they had dug through the snow that we had painstakingly shoveled into place, then chewed through the twine holding the bales together, and, in what must have been the most fun ever, they had shredded, chewed, rolled in, and scattered the straw far and wide. I never did get that perfect low-angle shot of Kamots howling with Williams Peak in the background. On the other hand, I gained a greater understanding of a wolf's boundless curiosity. So I consider that a success.

THIS ILL-FATED LANDSCAPING PROJECT was born in our imaginations and nurtured by our curiosity. What could we make? In what ways would it benefit the film? How would the wolves react to it? I wanted to find out, and although the results were not as I had anticipated, it was an illuminating and amusing discovery nonetheless. Curiosity had led us to try something we hadn't done before, to imagine the possible outcomes and to put the plan into action. The complexity of our undertaking may have been undeniably human, but the drives that set it into motion were not—at least not exclusively so. In fact, the reason it failed was because my subjects—the wolves of the Sawtooth Pack—were just as gripped by curiosity as I was. As we were building, they were wondering: What's this thing? Why does it smell different? What's it made of? Why are these two-legged creatures fussing over it? Is there anything worth finding inside of it? They wanted to find all this out.

Every day the Sawtooth Pack greeted their world with the same curious and slightly mischievous spirit. As much joy as this project brought me, I often awoke at wolf camp with a pit in my stomach, consumed by worries of the day. There were expiring permits to renew, suspicious ranchers to appease, and, looming above everything else, the search for a permanent home for the pack. I was consumed with worries, but not the wolves—how differently they behaved as they awoke on a winter morning. They'd rise from the little hollows of snow in which they'd slept; stand and shake off the night's ice from their fur; have a long, luxurious stretch; and happily greet each other. They seemed genuinely happy to find themselves in this world again and eager to embrace whatever the new day would bring.

Every new experience prompted joyous investigation. As winter tightened its grip, the little stream and pond that graced the wolves' home iced over. The wolves seemed to anticipate this event, eager to carry out what I liked to imagine were their own playful science experiments involving the states of matter. I first noticed their fascination with the frozen pond in their early years. In any season, the pond was one of their favorite places to play, so I was not surprised to see Kamots, Lakota, Matsi, Motomo, and Amani all gathered there one cold winter morning. They all seemed to be fixated on the pond's surface, and every so often one of them would bounce on his forepaws the way they would when catching mice in the meadow. I didn't want to disturb their activity, so I took my time getting close enough to see what they were up to. The surface of the pond had frozen fast but was still thin enough to be transparent. As I got closer, I realized that the wolves were chasing the little air bubbles that were being carried along under the ice. It occurred to me that the phenomenon of a solid but transparent object was altogether new to them.

This phase of their experiment was short-lived, because after a few pounces the ice broke and the tantalizing air bubbles instantly vanished. But then the wonder of shattering ice became their new object of investigation. I watched Motomo go through the process from start to finish, first cracking the ice with his paws and then bending down and gingerly picking up a thin piece of ice in his jaws. He appeared to find a sense of achievement in successfully picking up a large chunk and having it stay in one piece as he held it between his teeth, because he began prancing in a small circle, wagging his tail like a pup and showing off to the others. After enjoying that moment, he let his prize fall to the hard surface of the pond and shatter. He watched the little shards of ice skid in every direction. Each wolf conducted some variation of this experiment, getting particular pleasure out of the experience of taking a single solid object and smashing it into dozens of pieces.

WATCHING THE WOLVES engaged my own curiosity, and I think they felt the same way watching us. I often found myself wondering who was observing whom. Long after our ill-fated mini-mountain of straw had decomposed and Jamie had joined me at wolf camp, we began a more substantial construction project. I had been looking for a way to become a less obtrusive presence in the pack's life. Jamie's arrival, and her new energy, sparked the idea to move our camp inside the wolves' territory so we'd disturb them less as we entered and exited the enclosure and—we hoped— fade into the backdrop of their lives. But before that could happen, though, we had to do some very visible construction in the middle of their home.

The wolves greeted this week-long project with the same inquisitiveness as they did our other activities. By this time I had learned how their curiosity could easily turn to mischief whenever we weren't around to keep an eye on them. Still it was difficult to keep track of every tool and every bit of lumber.

Our new tented camp consisted of a raised platform upon which our big, round yurt was to stand. From the surrounding forest we cut dead lodgepole pines for the large primary support posts of the platform. We bolted smaller diagonal braces to reinforce the structure, then added a frame on top and nailed planking to it to form a deck. All the while the wolves observed our work. I knew the moment we turned our backs one of them would sneak in to seize a hammer or spill a box of nails, but I was surprised to discover they also took our natural building materials. The pack spent every day surrounded by the same lodgepole pines from which our support posts and braces had been created. The branches and logs that littered the ground failed to spark the slightest interest, but once a human had touched an object, it became something special. If something seemed important to us, it became fascinating to them, even if it was something as simple as the scrap from the end of a saw-cut log. It seemed as if they wanted to learn as much as they could about what we were doing.

Once the platform was built, we constructed a steep staircase. The steps were open, so you could see the ground below as you climbed to the yurt—a bit too precarious for a wolf to try coming up the stairs, we thought. Then Jamie and I began setting up the big canvas yurt on top of the platform. We planned to enclose the new camp in chain-link fencing before moving any food or personal belongings inside, so the wolves wouldn't steal our things. But during the earliest phase of construction, we had no barrier. One day, we were inside the yurt, building our storage and cooking area when a huge shadow appeared backlit on the canvas wall—like Little Red Riding Hood's worst nightmare. It made us both jump. Kamots, always the most brazen, had scaled the eight-foot staircase and was exploring the foreign structure.

I have since read that in cognitive experiments between wolves and domestic dogs, wolves scored higher on pure problem-solving ability, whereas domestic dogs scored higher on their ability to communicate their needs to a human and get help solving the task. It stands to reason: Dogs have evolved alongside the best problem solvers on the planet, so learning how to talk to us got them better results than struggling on their own. I don't think a dog could have climbed those precarious stairs without Jamie's or my showing it the way. But a wolf is a different beast.

To me the more interesting question isn't how Kamots climbed it, but why. The eight-foot stairway must have been a struggle for him. There was no food at the top, no tangible reward of any kind. I finally decided that he did it simply because he was curious. His desire to know what was up there was motivation enough. To complete his investigation, he stole a trash bag of building debris. After that, we quickly removed the first half dozen steps of the staircase to prevent further invasion.

Wolves are not the only smart, inquisitive creatures. Animals from primates to pigeons to octopuses have been studied for their ability to solve problems and manipulate their immediate environments. On the surface, human beings have more in common with other toolmaking great apes than with a canine species. But I feel we share a certain kind of curiosity more deeply with wolves than with any other creature on Earth. Wolves and humans are both great explorers.

Some 125,000 years ago, our African ancestors gazed across the Red Sea and wondered what was on the other side. As humans spread into Asia, we encountered *Canis lupus,* another smart, curious, expansive species. Gray wolves once populated most of the Northern Hemisphere because they, like us, were wondering, seeking, and moving.

As instinctual explorers, both wolves and humans possess the same inner contradictions. We are creatures of community and family. We have a strong sense of place, of home, and we'll defend our territory to the death. Yet within both our species burns a desire to break away and see new places. We want to know what lies just over the next mountain or around the next river bend. This may be yet one more reason why wolves and humans forged such tight bonds. As human beings pushed ever farther into new territory, wolves joined us on the adventure.

Like their forebears, wolves of today are seekers. In North America they are embarking on a new voyage of discovery into territory they once knew but lost more than half a century ago. As a handful of gray wolves were reintroduced into Yellowstone and Idaho, a few intrepid others were crossing the Canadian border into Montana. Some 2,000 more clung to the northernmost slivers of Minnesota and Michigan's Upper Peninsula. From these seed populations, wolves launched expeditions into Colorado, Utah, Washington, Oregon, Nevada, California, and Wisconsin.

A wolf seeker is embodied in a special type of individual called a disperser. Some wolves are content to spend their lives in the pack in which they were born. Dispersers are not. They can be male or female and are usually two or three years old—old enough to have reached adulthood but young enough to yearn for something new, something they can only imagine. Pulled by the unknown, dispersers set out alone with the ultimate goal of finding a mate and their own territory—driven, perhaps, by the simple joy of discovering someplace new. With only their four feet to propel them and no map, disperser wolves can travel staggering distances.

In 2009 a pup was born to a pioneering wolf pack that had set up home in Oregon. The pup's mother was a disperser—an adventurer in her own right. She had crossed the rapids of the Snake River from Idaho, found a mate, and settled along the Imnaha River in northeast Oregon. As the first in a new state, the Imnaha Pack was the subject of a great deal of study and public scrutiny. Fish and Wildlife officials fitted a few of them with tags and tracking collars so that their movements—and their impact on livestock—could be closely monitored.

By the time he was two years old the Oregon pup had

been captured, fitted with a GPS collar, and given the identification OR-7. Unfortunately, as a new pack in hard-line anti-wolf territory, the Imnahas were also subject to aggressive management and public hatred that led to the killing of several pack members by the Oregon Department of Fish and Wildlife. Perhaps the disruption and strife inflicted upon his family drove the young OR-7 to travel. Or perhaps he was impelled by his own internal desire. Whatever the reason, in September 2011, OR-7's GPS collar indicated that he had crossed the Cascade Mountains, thus becoming the first confirmed wolf on the West Coast of the lower 48 states since the last Oregon wolf was killed in 1947.

OR-7's journey had just begun. He turned southward, lingered in the Soda Mountain Wilderness, forged past Crater Lake and across the Klamath Basin. In late 2011 he made history again, becoming the first wolf to set foot in California in nearly a century. For the next three years, OR-7 flitted back and forth between California and Oregon until at last he found a mate and settled in the Rogue River Valley, near the California-Oregon border, to become the alpha of his own pack. All told he wandered well over 1,000 miles, and his adventure captivated the people of his adopted home state, many of whom followed his movements on a fan site. The conservation group Oregon Wild held a competition to name their new celebrity. In 2012, Oregon's school students voted to name him "Journey." As of 2017, Journey, a wise old wolf of eight years, remains the alpha of the Rogue Pack.

In many ways Journey was a lucky wolf. His travels took him through the sparsely populated Cascade Range, where he could wander in relative safety. The Imnaha Pack that stayed behind in ranching country was not so fortunate. Blamed for livestock loss, the last four Imnaha wolves were killed in 2016. In general, though, the modern landscape of America is an unforgiving place for predators with a thirst for travel. Many wolves, like those in Yellowstone, are relegated to small islands of protected land surrounded by cattle country and thousands of unfriendly humans. A wolf's natural urge to wander can lead it onto dangerous ground in very short order. Such was the fate of Yellowstone wolf Number 253.

Although he had an official identification, everyone knew him as either Limpy or Hoppy—his three-legged gait made him one of Yellowstone's most easily recognizable wolves. By the time wolf-spotters noticed him as a juvenile, he had already sustained the injury that left him partially crippled. Limpy was a young member of the famous Druid Peak Pack, born in 2000. Observers admired his tenacity on the hunt, keeping up with his parents and helping to take down elk without his front right leg ever touching the ground.

In 2002 young Limpy was struck with wanderlust. I find it remarkable that of all the Yellowstone wolves, the one with the hobbled foreleg decided to go on a long trek, limping 200 miles through western Wyoming, all the way to Morgan, Utah. I imagine him as the wolf incarnation of John Wesley Powell, who lost an arm in the Civil War and went on to lead the first expedition down the Colorado River, through some of the most unforgiving land in the country. Powell's notes from the journey convey that interplay of curiosity, fear, hope, and determination that propelled him forward:

> We have an unknown distance yet to run, an unknown river to explore. What falls there are, we know not; what rocks beset the channel, we know not; what walls ride over the river, we know not. Ah, well! We may conjecture many things.

Limpy had that kind of indomitable spirit, that yearning to seek out the unknown. As one reporter said, "His heart seemed stronger than his leg."

Limpy may also have been searching in vain for a female wolf, but it would be uncharacteristic for a wolf to turn his back on the highest population of available females in the country to look for a mate where no traces of his kind remained. It appears he really was gripped by the desire to explore. In any case, his stint as the first wolf in Utah in 70 years was short-lived. Of all things, one of his good legs found its way into a coyote trap. U.S. Fish and Wildlife officials collected him and released him at the northern edge of Grand Teton National Park. Instead of striking out on his own again, Limpy, now operating on two and a half legs, crossed the territories of several hostile wolf packs and returned to his old home and his birth pack. Back among the Druids of northern Yellowstone, he settled into the rank of beta wolf. Together he and the famous alpha pair, 21 and 42, became known as the Druids' "Big Three," largely responsible for the pack's incredible success.

The drive for wolves to stay in their family, their pack, is a strong one. In most wolves it is stronger than the urge to break off alone, but Limpy wasn't like most wolves. There's no exact count, but over his lifetime he must have covered well over a thousand miles. Even as he entered his eighth year, he was still given to wander. In the end, his spirit of exploration proved fatal.

In 2008, Limpy left his family and the protection of Yellowstone once more, and he headed southeast into Sublette County, Wyoming, where stores sell T-shirts depicting a wolf in crosshairs with the slogan "Smoke a Pack a Day!" For his entire life, Limpy had been a stalwart deer and elk hunter. No matter how deep into cattle country he roamed, and despite an injury that could drive him toward easy prey, he never attacked livestock. From a ranching standpoint, Limpy was a model wolf, but in Sublette County, someone shot him all the same.

Opponents of wolf recovery insistently stoke fears over the animal's insatiable hunger for sheep and cattle. The truth is, fewer livestock animals die as a result of wolf predation than by storms, injury, disease, or even attacks from other predators. But statistics don't seem to matter to those who cling to the 19th-century mind-set that the only good wolf is a dead wolf. When a wolf dies at the hands of man, it is almost never an official act of last resort. Killing wolves is a quick and easy appeasement to ranchers and hunters who didn't want them there in the first place. Changing livestock management practices and implementing nonlethal predator control requires cooperation, money, time, and political will, whereas half a dozen wolves can be shot from a helicopter in an afternoon.

Official wolf culling is just the tip of the iceberg. We also kill wolves recreationally, regardless of their impact on livestock, and we pass laws to minimize the penalties for illegal poaching and maximize the body count. We shoot and trap entire packs that choose to live in wilderness, far away from people and ranches. If wolves can't live in wilderness, where can they live? Even as we hold wolves to a near-impossible standard of behavior, most of them actually manage to live within our anthropocentric rules. But we shoot them anyway.

In Wyoming, people believe in the traditional values of the American West: bravery, independence, perseverance, and self-reliance. Yet I doubt that the hunter who shot Limpy, wolf #253M, took even a second to pause and realize that the creature he was about to kill was the embodiment of all the qualities he admired: an adventurous spirit, full of courage and curiosity. If he had, would he have pulled the trigger?

CHAPTER EIGHT

FIND COMPASSION

Jamie

It was the end of December when I arrived at wolf camp for the first time. Just a month earlier I had a job and was living in the suburbs of Washington, D.C. Now here I was, sitting next to Jim in his Volkswagen van as we wound our way over a mountain pass toward life in a tent without electricity or running water. I'd call that a leap of faith.

The closer I got to my destination, the less anxious I became. The land was so remote and unfamiliar, but the scenery was spectacular, with the jagged, snowcapped mountains towering over a broad, wide-open valley. We drove along the sparkling headwaters of the Salmon River northward. At the town of Stanley we took a sharp turn onto a dirt track across miles of sagebrush toward a dark forest of lodgepole pines and the foothills of the Sawtooth Mountains. As we steadily climbed in elevation, the dusting

Lakota

of snow became nearly a foot deep, so that we had to stop and hike the final mile into this new home of mine that Jim called wolf camp.

With our gear strapped to our backs, we left the scrubby sagebrush behind and entered an alpine glen with ice-crusted creeks and stands of pine, willow, and aspen. As we approached, I saw a small cluster of white tents peeking through the trees. I have to admit I was relieved to see that it wasn't the pup tent in the forest that I had feared. In fact, I was amazed at how substantial and organized the camp was. Jim and his small crew had created a comfortable outpost. It was basic, of course—just a couple of wall tents and a large circular yurt—but it looked cozy and secure.

Jim told me that whenever he returned to camp after being away for a few days, he liked to announce himself to the pack. He stopped and cupped his hands to his mouth and howled. For a few seconds there was only silence, then a strange sound began to rise from the forest behind camp. At first I thought it was a person calling back to Jim from a distance, but other voices quickly joined in. It literally stopped me in my tracks and left me standing in the snow, mouth agape. The wolves were howling in response. It was an otherworldly sound—joyous, mournful, calming, and exhilarating all at the same time. It wasn't at all like the clichéd wolf howls from the movies. At a distance, the sound seemed to come from everywhere, floating on air.

Had I not known only 5 wolves were present, I would have guessed that the number was closer to 15. I later learned that wolves purposefully avoid singing in unison. Each wolf varies its pitch to achieve a perfect dissonance, perhaps to make the group sound more numerous. Whatever their reasons, the effect is beautiful. I've heard wolves howl too many times to count since then, and that sound never stops thrilling me. But I'll remember that first time forever.

At first I didn't even try to settle in or unpack. I wanted to meet the wolves. Strangely, I found myself suddenly nervous, feeling like the new kid at school. They were waiting for us, whining in expectation, eager to greet their friend Jim and this stranger he had brought with him. I found myself worrying if they would like me. Would I be worthy of their trust?

We entered through a double gate, and the wolves began to gather around me excitedly. Jim suggested I crouch down so I wouldn't get knocked over. I did as instructed, and was immediately engulfed by a soft, fluffy tornado with tongues. Wolves tend to inspect anything new by sniffing and licking—and that included my face. I tried to keep my mouth shut, but the urge to smile was overwhelming. At one point one of the wolves even managed to get his lower canine tooth stuck up my nose. That was a bit disconcerting. Finally, having inspected me to their liking, the wolves moved on to other pursuits.

The only wolf I did not see on that first day was the one called Lakota, the pack omega. Jim assured me that I would meet this shy and special member of the pack when he was ready. Lakota was naturally shy around strangers, he explained—that part of his personality probably contributed to the other wolves pushing him into the omega position, low wolf in the hierarchy. Lakota probably also recognized that the other wolves would be excited by my arrival, and they tended to relieve their tension by picking on him.

That night Jim prepared dinner and we sat by candlelight in the cook tent, drinking wine. Now that I had finally met them, Jim began to tell me all about each wolf's personality: magnanimous Kamots; watchful Motomo; Jekyll-and-Hyde Amani; and, of course, meek and mild Lakota, whom I had yet to meet. All these stories made me wish I had been there from the beginning of the wolf project, yet at the same time they filled me with excitement as I thought about what lay ahead. At one point our conversation must have gotten a bit raucous, and the wolves started to howl, excited by our laughter. We let our dialogue drop and just shut our eyes and listened to them sing in the darkness.

THE FOLLOWING MORNING we still had not seen Lakota, and Jim and I began snowshoeing together, searching among the dense willows and fallen pines. I could see that Jim was a bit concerned, but he told me that the pack's territory was large and that he had sometimes searched for days for a wolf who wanted to be alone. The rest of the Sawtooth Pack was still interested in my unfamiliar presence, so for a while they followed us in single file like schoolchildren on a field trip. Eventually they lost interest in what we were doing, and one by one they wandered off to play or snooze. As soon as they were out of sight, we heard a rustling in the willows, and out crept Lakota. He had been following us the whole time.

One of the first things I noticed about him was his posture. He kept his tail tucked, his shoulders hunched, and his head lowered as he moved uncertainly toward me. It wasn't until he reached me that I realized he was a huge wolf, one of the biggest in the pack, but the way he carried himself made him look smaller.

Lakota approached me and timidly licked my face. I ran my hand down his back and through his new winter coat. His skin was riddled with small bumps and scabs where the other wolves had nipped him, and he had small scars on his muzzle where the fur would not grow back—signs of how a dominant wolf sometimes grabs the muzzle of a subordinate.

During my first few days at wolf camp, Lakota and I would find time to be together, meeting clandestinely in that same meadow where we first met. Gradually his trust grew and he began to relax. Then one day he took his paw, gently placed it on my shoulder, and gazed at me with his deep amber eyes. We sat that way for quite a while. From that moment on he would forever hold a special place in my heart.

I'VE ALWAYS FELT COMPASSION FOR underdogs and vulnerable creatures. I think that emotion is what attracted me to the job I had, caring for animals at the National Zoo. Once I joined Jim at wolf camp, this instinct of mine turned to a protective feeling toward the sweet and beleaguered Lakota. I could see how deeply he cared for the pack and how joyful he felt when they allowed him to start a lighthearted game of tag, and I could see his pain when one of them decided to bully him.

Pack rallies were an especially trying time for Lakota. A "rally" is the term biologists use for the high-energy gatherings that wolves engage in. Often wolves rally before they head off on a hunt, like a sports team psyching itself up before a game. Sometimes they occur for no reason that we can discern and appear to be celebrations of pack solidarity. Frequently the charged atmosphere of a rally provokes dominance displays in which one wolf asserts its authority over another. During such rallies Lakota would slink among the others and pay his respects to Kamots, with his tail tucked and his head nearly scraping the ground, but Kamots was seldom aggressive toward his brother. More often it was Amani who would rush at Lakota, snarling and flashing his teeth and making a big show of authority. Lakota would flip over and surrender. Although Lakota was rarely hurt in these displays, his cries were painful to hear. Amani would stand triumphantly over Lakota, making the omega beg to be let up. Once he was satisfied that his point was made—that he was the more dominant wolf—Amani would let up and Lakota would slink to the sidelines.

The pack rally usually culminated in a group howl, and then it was heartening to see Lakota joining in. The way a howl starts is a curious thing to behold. Sometimes it didn't seem entirely voluntary. It was as though the sound just welled up inside them and had to pour out whether or not they wanted it to. As the other wolves howled, I'd watch Lakota throw his head back tentatively and begin howling away. He had a beautiful voice, too, and I sometimes felt that he was singing the blues and letting his heart pour out, lamenting his unfortunate status.

Being the omega wasn't in Lakota's control; the other wolves had forced him into that position. All the same, wolves seem to be born with certain qualities that lend themselves to the role. No other position in the pack requires as much talent for diplomacy and appeasement. The better an omega is at coaxing the others into play and relaxing hierarchy for a while, the better his or her life will be. Lakota knew he was at the bottom of the pecking order, and in his way he was good at it. His role as omega was to keep the mood of the pack light, and he did it well. There was never any doubt that he belonged.

When the other wolves mistreated Lakota, my heart went out to him, but I could never intervene. The goal of our project was to document wolf behavior as naturally as possible, so we followed a strict code of interacting with the wolves only on their terms. We were responsible for giving them a safe home, regular food, and, if necessary, veterinary care. But in their day-to-day lives, we were committed to letting the wolves be wolves. It was part of the foundation of trust that we had with the pack. Sometimes that was difficult. If two wolves had a dominance scuffle, all we could do was watch and hope all would end well.

When abuse from Amani or one of the others became too great, Lakota would simply retreat into the willows to be by himself. It was heartbreaking to watch. In my earliest days at wolf camp, it seemed to me that Lakota lived in a friendless world and that those moments we spent together might have been his only relief. But as I got to know the pack better I discovered, to my joy, that this was hardly the case. Another wolf understood his timidity and felt for him as I did.

Wolf communication isn't always easy for us to read. But we had the advantage of capturing so much of it on film, much of it shot in slow motion or in a sequence of rapid-fire transparencies, so we could go back and observe details even more closely that we might have missed in the moment. At times, we weren't able to decode the wolves' behavior and understand what was actually going on between them until we looked at our images or watched the film footage. In one particular sequence of images of a pack rally, for example, we noticed a pattern emerging—and Matsi's reputation as a peacemaker was born.

THE RALLY HAD TURNED SOUR on Lakota, as occasionally happened. Even though it would have been safer to keep to himself, Lakota was part of the Sawtooth Pack, and when they howled together he wanted to join in. In this particular rally, Lakota wound up being the target of the extra energy that permeated the event. The pack had formed a mob around the hapless omega, and a few of them were displaying their dominance by nipping and snarling and standing over him. Even the yearling Chemukh got involved.

As always, Amani was the main perpetrator. He just didn't seem secure in his position as a mid-ranking wolf, and he seemed to feel that the best way to avoid winding up as the omega himself was to do everything he could to keep Lakota down. In the midst of this rally, he literally climbed on Lakota's back and started nipping at his neck. Lakota sank to the ground and tried to worm-crawl his way out of the frenzy.

A moment later, Matsi charged into the fray. When I watched this unfold in real time, I assumed that Matsi was simply showing dominance over Lakota, as the others were. Looking at the sequence of rapid-fire images, however, I could see where Matsi's energy was actually directed. He wasn't digging into Lakota; he was body-checking Amani out of the way and giving Lakota the chance to get away. Matsi was disciplining Amani for disciplining Lakota.

At first we thought this was some kind of fluke, but we began to notice this kind of behavior more and more. At mealtime Kamots enforced the hierarchy and sometimes prevented others from eating for a while, but once he made his point he usually mellowed out. Often it was Amani, not Kamots, who tried to keep Lakota from eating. On one occasion the entire pack, except Lakota, was enjoying a big elk carcass. Lakota put his head down and crawled forward, trying to look as inconspicuous as possible but still getting a place at the table. Amani snarled and grabbed Lakota by the neck. Lakota let out a cry and flipped onto his back. The next thing we knew, Matsi left his place and lunged toward Amani, who yelped and ran a few yards away. Matsi didn't pursue. He just walked back to his spot on the carcass. His message was clear: "Leave him alone." The entire scuffle lasted less than 10 seconds, and in the end Amani and Lakota were both eating without trouble.

We were never able to explain the special bond that Matsi developed with Lakota. Their relationship was particularly surprising because Matsi had participated in pushing Lakota into the omega position to begin with. Once the ranks were established, however, Matsi was extremely gentle with the omega. Even though we couldn't explain it, we did have a simple word for it. Matsi and Lakota were friends.

As adults, the two often slept alongside each other. At the very least, Lakota knew Matsi was the one wolf who would not pick on him and might even defend him against others, but there seemed to be more to it. When Matsi went exploring, Lakota frequently joined him, and Matsi genuinely seemed to enjoy the company. Being with Matsi gave Lakota the freedom to do things he wouldn't dare do around the others. He felt relaxed enough to roughhouse, jumping on Matsi's back for example—he would never have invited Amani or Motomo to play in the same way. Matsi relished these invitations and launched into playful tumble or a game of chase with Lakota. It was such a delight to watch these two frolicking together and to see the look of joy on Lakota's face, freed for a moment from his burdensome role as the underdog. I believe that Matsi genuinely understood how much

Lakota suffered as the omega and wanted to offer him some relief.

IF ONLY IT WERE THAT SIMPLE! I guess all good friendships can be tested, and there was one puzzling interaction between these two friends that we love to tell wolf biologists about. One day, as Jim and I were photographing the pack, we noticed Lakota cowering under a tree. He was tucked into a deep crouch with his ears plastered back, looking as if he were trying to be absorbed by the ground and disappear. As I watched him and wondered what was going on, Matsi walked up to Lakota, lifted his leg, and urinated all over the unfortunate wolf's back. I had missed whatever earlier altercation had provoked this extreme demonstration of dominance, but it was clear that Matsi felt the need to put Lakota severely in his place. But apparently neither of them held a grudge: A few hours later I spied the two playing by the pond together, and that night they slept side by side. Whatever the disagreement, it seems that wolves can forgive.

Forgiveness, compassion, and empathy are very complex emotions. First they require having a concept of one's own individual self and of others as individuals as well. Then one has to make the mental leap of imagining oneself in the position of the other. Matsi must have understood how he would have felt if he were in Lakota's predicament. Otherwise, why would he have gone out of his way to help Lakota? It's not an intellectual process; it's a gut feeling. I believe Matsi could feel what Lakota was going through.

There's no doubt that the wolves understand their own individuality and the individuality of others. We saw evidence of this in the Sawtooth Pack every day. There's a story that Jim likes to tell of an episode that happened before I arrived at wolf camp. It was the first snowfall of the year, and it sent Kamots, nearly two years old, charging around with excitement. He ran back and forth, not playing tag or chasing another wolf, just darting about full of energy and exuberance.

Amani, less than a year old and not yet full size, wasn't quite sure what to make of the alpha carrying on this way. He sat down in the fresh snow and watched as Kamots burst out from the willows at full speed. Then Kamots, gleefully out of control, barreled right over Amani. Amani did a cartwheel and sat up, blinking and covered in powder, not sure what had just happened. Kamots slammed on the brakes, turned, and hurried back to Amani, sniffing the little wolf up and down and giving him a reassuring lick. It was a touching gesture, and it reveals Kamots's mind-set clearly. He understood that he was a big wolf and that Amani was smaller and weaker. He realized that the collision might have been upsetting to the younger wolf, and he cared enough to stop what he was doing to go check on his little packmate.

Observations such as this aren't unique to the Sawtooth Pack. On a visit to Fairbanks, Alaska, a fish and game biologist showed us the skull of an average-size male wolf that bore unmistakable evidence of a broken jaw that had healed. A defensive moose or caribou probably dealt him that injury while he was hunting. It's not uncommon. What is more interesting, though, is that the jawbone healed and the wolf went on to live for several more years. Such an injury would have rendered the wolf unable to tear chunks of meat from a kill and possibly even unable to chew. The only way he could have survived this injury is for the other wolves of his pack to feed him. At the least they would have had to bring him pieces of meat and likely even regurgitate partially digested food for him, as they would a pup. The only way that wolf could have survived with a broken jaw was through the generosity of others.

We can only guess the motivation of this wolf's packmates. I suppose there's a chance that nursing a fellow wolf back to health could be a purely selfish act. Maybe he was a good hunter and the others wanted to keep him around. But I just don't see them being motivated by such cold pragmatism. After seeing so many moments of unambiguous kindness between the wolves of the Sawtooth Pack, I'm sure this was an act of compassion between family members. Is it really surprising? When we humans see a friend or relative in pain, we feel an emotional urge to ease the burden.

Recently there have been some fascinating studies into the connection between human empathy and a weird social phenomenon that most of us have experienced—contagious yawning. I can remember sitting in a college classroom, bored by a lecture. One student would yawn and then the next, and the yawn would pass from student to student like a wave across the room. We'd notice and begin to titter, much to the professor's great annoyance. It turns out this didn't mean we were all equally tired or bored, but rather that we were fairly healthy mentally.

Yawning in response to another's yawn isn't very well understood, but as far as we know, it has less to do with breathing and more to do with sociability. It's an urge connected to our own emotional cognition—that is, the ability to tune in to the emotions of someone nearby. Not surprisingly the earliest studies of nonhuman contagious yawning focused on other social primates. Chimpanzees, bonobos, and gelada baboons all experience it. It has since been observed that humans can sometimes pass a yawn to a dog, but it was suggested that this was because dogs have spent thousands of years evolving in our company. We saw the Sawtooth wolves pass yawns from one to another all the time. Quite often Jim and I would be watching them, he with his film camera and me with my audio equipment, waiting for them to do something worthy of rolling precious film. As they lounged together, one of them would yawn and it would break like a wave over the entire pack. Before I knew it, I would feel the uncontrollable urge take over, then I'd look over at Jim and he'd be yawning too. We thought it was funny, but we never assigned any significance to it. Now it's been scientifically confirmed that wolves also experience contagious yawning—an indication that they, too, have emotional cognition, the roots of empathy.

It is enough to realize that the capacity for empathy and compassion, although undeniably human, is not uniquely human. But for me the lesson of the Sawtooth Pack didn't stop there. Compassion was their first impulse. It was instinctive in them just as it is in us. But wolves aren't perfect either. Amani especially could forget his instinct for compassion and become a bully. The reason, I believe, is that his own insecurities about his position in the pack hierarchy sometimes got the better of his inclination to care for his packmates.

Every wolf feels itself pulled by opposing forces. On one side there is ambition, self-interest, and fear; on the other there is forgiveness, compassion, and empathy. That's what makes wolves so endlessly fascinating. When you think about it, that's what makes human life interesting too, but wolves are less prone, or less able, to blur intentions and counterfeit emotions. As a wolf struggles to find a balance between these two opposing drives, its motivations are laid bare. In the end there is no contest. It must follow the path of compassion, of cooperation, of togetherness above all. Survival offers no other choice.

CHAPTER NINE

CHERISH ONE ANOTHER

JIM

A MISTY RAIN GATHERED ON MY WINDSHIELD as I raced toward wolf camp, pushing the limits of my old van on the windy road over Galena Summit. I had been away from camp for the night, caring for the second litter of Sawtooth pups—Motomo, Amani, and Matsi—who had to go through a round of vet check-ups before they could join their new pack. The crew member who was in charge of the pack while I was away had driven the muddy jeep track to the town of Stanley after dark to call me from a pay phone. One of the wolves was missing.

The wolf was a sibling to Kamots and Lakota, a mysterious and shy female with a coat of black and brown and piercing yellow eyes. We had given her the name Motaki, the Blackfoot word for "shadow." She was the pack's original omega in the earliest years, before Lakota

Matsi and Kamots

took on that role and before Jamie joined my life at wolf camp.

I had bonded with all the wolves of the Sawtooth Pack, but Motaki had a special hold on me. Other wolves may feel the biological pull to be an alpha, but Motaki didn't have an ambitious bone in her body. She avoided the minor confrontations and dominance displays that are part of life in a wolf pack. As sometimes happens in human society, her sensitive spirit relegated her to the bottom of the social order, and she accepted that. As the original omega, Motaki was the first wolf to show me how important that role was within the wolf pack. I could see how her gentle nature helped soothe tensions and create harmony. She was always the one who could get a game of tag going and put every other wolf in a playful mood. Although she sometimes had to endure dominance and aggression, she howled with the rest of the pack, slept with the others, and always joined them at feeding time. There was never any doubt that she belonged.

Like Lakota after her, Motaki sometimes spent hours away from the rest of the pack, retreating to the denser forest to be alone. I had learned not to worry if I didn't see her for a while. This time, however, she had failed to show up when the field assistant brought in food, and that was not like her.

I arrived at camp and began searching along with the crew. The first thing we did was walk the perimeter of the enclosed territory to make sure a fallen tree hadn't brought down a section of the fence. Finding the fence intact, I doubted that Motaki had left the enclosure. We had designed it with an inward overhang at all corners so that even the most athletic wolf wouldn't be able to climb over it. A six-foot apron prevented them from digging under it. There was even a solar-powered electric wire added to the fencing for good measure, and there was no sign it had stopped functioning. The realization that Motaki was most likely still inside did not bring any sense of relief. It meant that she was close but unable to join the others. So we moved in from the fence line and began to crisscross through the aspens, willows, and evergreens where she might be hiding.

I found her in an aspen grove at the southeast end of the enclosure, lying still and silent among half-fallen trees. Across her belly ran a deep gash. The gentlest, most playful wolf whom I had known was dead.

My first thought, one I am somewhat ashamed to admit now, was to wonder if the other wolves had done it. I just couldn't think of any other explanation. But the more I looked, the more I could see that the pieces didn't fit. I had never seen her enter into a violent confrontation with the rest of the pack, but if she had this once, her legs and hind-quarters would have borne the scars that wolves deliver in these fights.

When I inspected the wound, I could see that something had started to eat her and then stopped. The other wolves? Again, it didn't add up. This might occur under the most desperate circumstances, but not in a healthy pack with a steady food supply. Plus, wolves eat by tearing at a carcass and ripping off sections, and Motaki's body was still intact.

Something about the way she looked brought a memory back to life. Years earlier I had documented the life of a mountain lion for my film *Cougar: Ghost of the Rockies*. I remembered how the cat I filmed would always lick and clean away the fur before beginning to eat her prey. When she ate, she always opened up the belly first. The pieces started fitting together.

With a picture of a new culprit in mind, I began searching for clues. Next to where she lay was a partially toppled aspen tree. Where the trunk forked, seven feet off the ground, I found black wolf fur. At the base of this tree were the dull claw marks of wolves, as if they had been trying to get at something up in the tree. Months earlier, I'd seen them claw at the base of another tree beneath a woodpecker's nest in a vain attempt to reach the bird. These new marks looked exactly the same. I looked around for the paw prints of wolves or any other animal but found nothing. It had been dry before the incident and had rained since, so the chances of finding prints were not good.

Nevertheless, I began to piece together a picture of what most likely had happened. The cougar, moving through his territory, had likely climbed a tree close to the fence and jumped in. Mountain lions and wolves have been known to kill one another, competing for territory. It was possible that the lion had caught wind of the wolves and taken them as a threat. The cougar had probably caught Motaki by herself while the rest of the pack was far away. The kill would have been swift and silent. The single crew member at camp was too far away to hear any commotion. The wolves may have caught up with the cougar, driving him to drag Motaki's small body into the tree, but eventually the cougar let go of his prey and dashed back over the fence to escape the pack. A fence containing a wolf would barely have slowed a powerful mountain lion.

FOR A WHILE I WAS SO WRAPPED UP in my own feelings about the loss of Motaki that I didn't consider the feelings of her own family and most loyal friends. A wolf's grief may not be obvious, but as I observed them I began to see the subtle signs, familiar to me because it's so much like the behavior of grieving humans. As anyone who studies them will confirm, wolves usually engage in some type of play every day. This was certainly true for the Sawtooth Pack. Yet after Motaki's death, a full six weeks went by during which I observed no play whatsoever. Motaki had been the main instigator of play among them, and without her, they just lost the spirit.

Normally when the wolves gathered for a pack rally,

they would howl enthusiastically together, often punctuating their song with excited yips, whines, and happy Chewbacca vocalizations as they whirled about each other in a display of camaraderie, pressing shoulder to shoulder. Following Motaki's death, their howling took on a different tune. They still howled, but not as a group. It was a long, slow howl, the kind of howl a lone wolf might make when it is searching for a lost pack member, not the lively rally sounds I was used to hearing. On several occasions, Kamots and Lakota didn't even bother to stand up to howl. The arrival of food still generated some interest, but other than that they seemed to drift about in a listless manner. Head lowered, eyes down, ears back, their behavior was not unlike that of a dog that has lost a human companion. There's only one word I would use to describe it: grieving.

After six weeks the wolves began to return to some normalcy. Introducing the second litter of Matsi, Motomo, and Amani gave them something to focus on and maybe even a sense of hope. Yet months after Motaki's death, I observed some of the most moving behavior I'd ever seen from the pack. I was walking among the aspens with the pack following curiously behind me, as they sometimes did, when I reached the spot where I had found Motaki's body. As the wolves caught up, they broke away from me and inspected the area, silently sniffing the ground. At that moment every one of them took on the posture of a submissive omega.

Even proud Kamots had his ears pinned back and his tail drooping low. There's no better way to explain it than they were depressed.

I would have given anything to know what was going through their heads as they did this. How much did they comprehend her absence? How deep was their understanding of death? At this moment they appeared to grasp what had happened at that exact location many weeks before, and they were mourning the loss. Motaki had been their littermate, sister, playmate, and packmate. She was a cherished member of their family, and I believe they missed her.

Perhaps the wolves even remembered trying and failing to help her. That's where my speculation ends. I have no doubt that they felt the pain of losing her, but I'm fairly certain wolves don't experience regret or self-recrimination. Humans certainly do. Once I realized what had happened to Motaki, I doubled my own agony with a hundred "should have"s and "if only"s. In some ways I admired the wolves for their ability to experience grief in its pure form, feeling the loss in all its intensity and then moving forward.

Among my regrets, analysis, anger, and blame was a realization that I didn't fully comprehend how deeply they cherished Motaki until she was gone. The project was only in its first year. The behavior I had witnessed had always simply been "wolf behavior" to me. I couldn't quite understand that within all of their interactions—the eye contact, the gentle games, the pressing of shoulders while walking—were countless expressions of affection. I came to understand these things so fully, so completely only when faced with their absence.

When Motaki died, something began to change in me. Until then I had thought of myself as a wildlife filmmaker. I had always assumed that after the wolf project I would begin a new film on a new animal subject. But seeing the wolves grieve and struggle with the loss, and seeing them shower so much affection and care on the second litter of pups, I realized I wasn't just a filmmaker anymore. My work, and all the energy and passion I put into it, was for wolves now.

GORDON HABER WAS ONE OF THE WISEST BIOLOGISTS I've ever known. He was one of the first to write unabashedly about the devotion wolves have for one another. He began to notice that courting wolf pairs snuggled while walking and lying together. He reasonably surmised that this has the same value as it does among human couples—a reaffirmation of their special bond. He especially noted that wolf pairs maintained this physical closeness and displayed "emotional attachment" well outside of mating season. Just as my clearest understanding of wolf family bonds came at Motaki's untimely death, Gordon's most moving observation of the devotion between a pair of wolves came at the moment when that bond was shattered.

The male and female were one of the last alpha pairs of the Toklat Pack that Gordon observed in Alaska. They paired in 2002 and together endured the loss of their first pups to a marauding bear, but by the end of 2004 they had successfully raised two litters. Most of the Toklat's territory fell within the huge protected area of Denali National Park and Preserve, but some of it fell outside the park boundaries. The hatred of wolves burns just as hot in many parts of Alaska as it does in the lower 48. Some of the most vitriolic anti-wolf people run legal traplines just outside the borders of the Denali park. Wolves, unable to discern whether they're inside the park or out, run into a gantlet of death the moment they cross the invisible boundary.

One of these leghold traps just outside the northeastern boundary of Denali ensnared the Toklat alpha female in January 2005, just before mating season. Leghold traps don't kill outright; they simply hold the wolf in place to starve and suffer from a painful wound. If a wolf is lucky, the trapper will appear within a few days to finish the job. The Toklat female languished in the trap for two weeks before she was finally shot. During her entire ordeal, her mate stayed in the area, most likely by her side. It's quite possible he brought food to her, desperate to help but ultimately unable to keep her alive. When at last she was killed, he and their young fled the area.

The loyalty he displayed during that ordeal would have taught us enough, but his behavior over the next few weeks would prove even more heartbreaking. When he and his offspring left the trap site, they crossed back into the national park and returned to the den site where most of the pups had been born. There the male dug through the snow and cleaned out the den, removing the leaves and loose soil and readying it for a new litter of pups that he would never father. The next day he traveled 14 miles back to the spot where she had been trapped, moving at such a fast pace that his pack often struggled to keep up. He searched for his mate with an intensity that Gordon called "almost obsessive."

He probably didn't see his mate die. When the trapper appeared, fear likely overwhelmed his senses and he fled into the forest. Perhaps he heard the gunshot and understood what it meant, but perhaps not. Most likely all he really understood was that his mate was in trouble, and then she was gone. Did he prepare the den and search for her, believing she was still alive? Or did he do those things because he was distraught to the point of madness, unable to accept what he knew?

When you look at wolves only as vermin, or as identification numbers, nothing can explain his behavior. Only through the lens of our own human empathy does it become clear. He and his mate had begun the process of pair bonding after they had first met. They flirted and snuggled and affirmed their affection for each other in a way that any human would recognize. They had suffered the loss of their first litter and persevered through the ordeal together. They worked tirelessly as a team to raise their subsequent litters. Powerless to help as she died, what else would he be feeling? The Toklat pair was as closely bonded as any human couple. They expressed their bond with a purity and openness that only the most enlightened humans would achieve. When he lost her, his grief was equally pure and terribly sharp. When Gordon left him in mid-February, he was standing on a high plateau howling over and over in the direction of the trapping site.

How far would this wolf have gone to help his mate if he were able? As humans we like to think that altruism and sacrifice belong to us alone. After all, it's not an easy mental or emotional process. The act of risking personal harm on behalf of another requires a complex cognitive analysis to take place before the ultimate moral decision. First one has to be able to foresee an undesirable outcome for another. Then one must deduce that putting oneself in danger might deflect harm away from others—and one must decide that saving oneself is not as important as saving the other, whether an individual or a tribe. We humans are inclined to claim all those cognitive processes and emotions for ourselves alone, but those who spend their lives observing wolves have observed exactly this type of behavior—the ultimate display of devotion and sacrifice.

RECENTLY JAMIE AND I were in Yellowstone National Park, talking to Rick McIntyre, who has been monitoring the park's wolves since they were first reintroduced there in 1995. At this last meeting he told us the incredible story of the Lamar Canyon Pack alphas.

The Lamar Canyon Pack had struggled over the years. Hunters had killed the pack's original alpha male and female. The next alpha male, a bruiser of a wolf known as Big Gray, took over the pack with a new alpha female, only to have it dissolve. As the rest of the pack dispersed, one subordinate female, called simply the Black Female, stuck by Big Gray. Eventually she became his mate and the alpha of the reborn Lamar Canyon Pack. Together they produced six pups in 2014, and held on to a territory in the northeast corner of Yellowstone.

As one of the few places in the lower 48 states where wolves are protected from hunting and trapping, Yellowstone now has a dense wolf population. As a result, territorial conflicts between packs appear to happen more frequently than elsewhere. As wolves move to follow elk herds, there is a high probability that they will cross into another pack's territory, possibly sparking a conflict.

In March 2015, the Lamar Canyon Pack did just that. With their six yearlings in tow, they set out westward on a hunting trip. The Black Female was pregnant with their next litter. Wolves seem to like to ramble in the early months of a new year, before their pups are born. They probably know that a period of restricted travel and responsibility is coming and want to have one last hurrah before settling into their den site. Once they return to their territory, the alpha female begins the process of clearing out last year's den and readying it for the new pups.

The Lamar Canyon Pack's journey took them across the territory of the larger, rival Prospect Peak Pack. Outbound they had crossed without incident, but they were not so fortunate on the journey home. As they attempted to cross, 12 Prospect Peak wolves descended on them from the east.

The pregnant black female and the six yearlings made a beeline westward, backtracking. In her condition, though, she could neither fight nor outrun the Prospect Peak wolves. The yearlings were barely older than pups. That left Big Gray as the only wolf of the pack capable of confronting the attackers. As 12 Prospect Peak wolves charged up a hill, he turned to face them. Rick McIntyre watched it all, and, as he said, "The action was deliberate, taking courage and resolve to have your enemy run straight at you while you stand your ground."

As the Prospect Peak wolves got close, Big Gray ran straight downhill back through them. It happened so fast that the Prospect Peak Pack had to turn around to go after him, giving Big Gray's mate and the yearlings more time to escape. The yearlings watched as the Prospect Peak wolves caught up with their father and fell upon him in battle. At their age they had neither the muscle of full-grown adults nor the experience to fight a battle like this. Still, they could run. One of the yearlings ran back into plain view, and when the Prospect Peak wolves saw him, a few peeled away from Big Gray to give chase, but they couldn't catch him. A second and then a third yearling did the same, drawing off more attackers. Big Gray was left in a four-against-one battle. Rick says he watched them fight for about five minutes until they disappeared into deep sagebrush. He assumed Big Gray had not survived the assault.

Soon Rick began to hear howling in the distance. The Lamar Canyon yearlings and their mother were calling for Big Gray. The Prospect Peak Pack, also hearing the howls, went looking for the Lamar Canyon wolves, but they couldn't find them.

Eventually, the Prospect Peak Pack gave up their search and returned to the site where they had left Big Gray, but he had disappeared—and they lost interest. But one young Prospect Peak wolf picked up Big Gray's trail and followed it, only to find that the Lamar alpha had a lot of fight left. Without 12-to-1 odds, the young Prospect wolf turned tail and ran. Big Gray, bleeding and gravely wounded, disappeared into the sage.

From this point on, all Rick knows is what he was able to ascertain by monitoring the signals from Big Gray and the Black Female's GPS tracking collars and following their movements. The next day he listened as the Black Female's signal showed her returning toward the danger zone, inching closer and closer toward her mate as he lay there, weak and unable to move. By nightfall the two signals were side by side. She had come to say goodbye.

A GPS collar is programmed to give off a special signal if it remains motionless for a certain length of time—this is called a mortality signal. The next morning Big Gray's collar began emitting that signal. The signal from the Black Female's collar showed that she had moved safely back into her territory at her den site. Rick believes that she had made a deliberate decision to visit her mate one last time. "The Black Female had to find her mate," Rick went on, emotionally, affected by his own story. "She had to let him know that it was not in vain," he says. "She was safe, his unborn pups were safe, and his yearlings were safe."

To follow the ever changing fortunes of Yellowstone wolves is to witness a real-life saga, complete with acts of heroism, sacrifice, tragedy, and resilience. Objectively, of course, there are no real heroes and no real villains. Wolves do what they need to do to survive, and they are always full of surprises. The Lamar Canyon Pack saga continued beyond this epic moment. The Black Female found herself in dire straits with a new litter of pups and six inexperienced one-year-olds, still needing support. So she accepted the advances of a wolf called Twin—none other than one of the Prospect Peak males who had attacked her mate only weeks earlier. Ultimately three other Prospect Peak males left their pack and joined them in the Lamar Canyon territory. In true wolf fashion, they helped raise Big Gray's pups. His legacy lives on.

SO MUCH OF WHAT WE HEAR about the devotion wolves have for one another comes from observations of bonded pairs in epic tales we've heard from Yellowstone and Denali. After Chemukh joined the Sawtooth Pack and grew to maturity, we were able to witness this kind of pair bonding firsthand between her and Kamots. Once nature had ordained that they become the alpha pair, their relationship became stronger and stronger. Kamots would make sure that his mate had a secure place at every meal. He was emphatic about this: There would be no squabbling; Chemukh would eat first with him. It was a complete reversal from the way he had treated her when she was a pup.

Over the years we saw again and again how every wolf in the family cared deeply about every other wolf in the family. In the beginning it was evident in the way the young pack reacted to Motaki's death. Throughout the project it was the special friendship between Lakota and Matsi, as Matsi protected the omega. In the end it was the ever endearing bond between two siblings. When Wyakin died, her devoted brother, Wahots, was found posted by her side.

For me, the moment that truly drove home the depth and power of the Sawtooth Pack's devotion to one another was one of the last moments Jamie and I would spend as their constant companions. I had always known the wolves could not stay forever in the idyllic territory beneath the Sawtooth Mountains. Our land use permits expired, and the Forest Service would not issue another extension. After a great deal of negotiation we reached an agreement with the Nez Perce Tribe to build a new home for the wolves in northern Idaho. It was a huge relief, but it also meant our time with the pack was coming to an end.

The move from the Sawtooth Mountains to the Nez Perce Reservation in August 1996 was probably the most stressful event of these wolves' lives, and of ours too. We did our best to ease their transition, transporting them overnight when it was cool, but I can only imagine how bewildering it must have been.

We waited until the heat of the day had nearly passed, and in a painstakingly coordinated procedure, fed each wolf a piece of meat that contained a sedative, reinforced with a mild injection. We covered their eyes and gently placed them in the crates. There was no need to do this to the pups, Ayet, Motaki, and Piyip. They were still so young that we just picked them up and placed them in the crate.

Kamots did not want to go under. Perhaps knowing the safety of the pack rested on his shoulders, he resisted the tug of sleep. Watching our trusted friend stumble, fight to his feet, and finally succumb to the sedative was pure agony. Even worse, as the drug began to wear off and the wolves awoke safely in their crates, Kamots did not rise. We had given him a fraction of the dosage that government veterinarians routinely administer to wild wolves, but his heartbeat was faint and his breathing shallow. More than any of the other wolves, the pack needed him desperately.

It seemed to take forever, but finally his ears twitched and his eyelids fluttered. We could see he was conscious but unable to rise. His yellow eyes, usually so bright and alert, seemed to be gazing at something invisible and far away. I reached into his crate and put my hand on his neck. "I'm sorry," I told him. "I'm so sorry. It's going to be OK, I promise." Jamie knelt beside me and whispered, "Kamots, please wake up. We need you. We can't do this without you."

Jamie and I agreed that the best course would be to make the trip and end the ordeal. We carefully hoisted the crates into a horse trailer, and as the sun began to drop into the Sawtooths, we pulled out across the flats toward Stanley and the north. We drove in silence, Jamie and I in our van, following behind the trailer, thinking of Kamots inside. Finally, I signaled for our caravan to pull over. Fearing the worst, I released the latch and carefully swung open the door of the trailer. Kamots was sitting upright in his crate, blinking. I put my hand up to the door of the crate, and he gently licked my fingers. A collective sigh went up from our small group. He was a bit groggy, but he was going to be fine. His eyes caught the moonlight as he looked at me, and I could see the old, confident Kamots returning. There was so much trust in his gaze. I wished I could have explained to him the reasons for this ordeal we were putting all of them through, and how painful this was for me too, but his look reassured me.

Just after 4 a.m. we arrived on Nez Perce tribal lands to the site of the new wolf enclosure. Our headlights spooked a flock of wild turkeys, which flew off with a ruckus. Those turkeys are about to have a rude awakening, I thought. It was still dark, so we settled down to sleep in the vehicle and let the wolves rest until sunrise. The elevation was considerably lower at the new home. I could feel the thickness and humidity of the air, and I supposed the wolves could feel it too.

After sunrise we carried the crates into the enclosure and lined them up side by side in a grassy meadow. We thought that it would comfort the adults to see the pups free and safe—it would give them something on which to focus

besides the discomforts of travel and the strange new surroundings—and so we let the pups out first. Piyip, Ayet,
and Motaki bounded out of the crate and into the grass.
Still at the age of blissful ignorance, they were untroubled
by the trip and eager to explore all the new sights and
smells. And then they turned back to see where everybody
else was.

Kamots was the first adult to be released. For a moment
he held his head low, tucked his tail, and looked around
tentatively. It was a posture of uncertainty—I think the
only other time I saw him carry himself this way was after
his sister, the elder Motaki, had died. I could see just how
stressful this was, even for him. Still, he went straight to
the pups and inspected each one. They gathered around
their father in a frenzy of licks and whines, overjoyed to
see him again.

We let Matsi out next. He enthusiastically greeted the
pups and then lowered himself slightly and licked up
toward Kamots. As quickly as we could, we released
the others, Chemukh, Motomo, and Amani, Wahots
and Wyakin. Each wolf went first to the pups and
then to Kamots. It was similar to their customary
morning greeting but with considerably more whining,
licking, and reassurance. They obviously had a lot to
talk about.

We purposely let Lakota out last. If he had been able to
greet the pups and Kamots before the others had a chance
to, they might have used him as an outlet for their stress.

We opened the door of Lakota's crate, but he held back. The pack had begun to move around, sniffing the ground with growing curiosity, confident enough to begin exploring their new territory. Still Lakota did not emerge. Kamots must have noticed. He stopped his explorations, turned, and walked back to stand beside Lakota's crate. The two whined back and forth for several seconds. At last Lakota, wide-eyed, poked his head out of the crate and took one tentative step forward. At that, Kamots stepped up and pressed his shoulder against his brother's. Lakota lifted his head, and they walked into their new home side by side.

Watching that scene, I could tell that Kamots would never leave his brother behind, no matter how stressful the situation was. He understood that the wary Lakota needed a little encouragement, and he knew how gently to coax him forward. It seemed fitting that, as Jamie and I prepared to say goodbye to the pack, they would make such a profound gesture of their devotion to one another.

Later that afternoon we gathered in a field, a few hundred yards from the enclosure, where the Nez Perce Tribe conducted a small ceremony to welcome the wolves. Amid the sound of drumbeats and voices we began to hear a distant harmony rising up into the air. The wolves, responding to the music, had begun to howl. I remember looking at Jamie holding her boom microphone, recording audio of the event with tears streaming down her cheeks. The pack continued to howl after the human song ended, and everyone stood motionless and listened until the last sound drifted away.

SEVERAL YEARS LATER, when we received word that Kamots had died, we were deeply saddened. The life of a wolf is so short, and he had meant so much to us. Kamots was the strong leader that the project needed, and I really believe that he was the one who held the project together through all its years. In Kamots we saw the power and wisdom that gave the other wolves confidence and security.

Kamots taught me so much about the companionship and trust that wolves have for each other. He showed me that if one is lucky, a wolf may extend that honor to a human as well. Losing an animal friend like Kamots is doubly painful, for I can never know if he realized how much he meant to Jamie and me. Nor did he know that as we went on to tell his story, he and his packmates are known worldwide as wolf ambassadors, sharing their truth with everyone. This is our greatest consolation: that people are starting to understand wolves better and to know them as deeply emotional, caring, and family-oriented animals that deserve the chance to survive.

For three weeks after Kamots's death, a single wolf was heard howling alone in the night. Knowing them as we did, we figured it was probably his brother, Lakota, grieving the loss of his magnanimous brother and friend, and letting the world know.

CHAPTER TEN

THE WOLF IN
THE MIRROR

JAMIE

I MISS WINTER NIGHTS MOST OF ALL, especially during those happy years when the eight wolves we knew best had formed a cohesive pack—at those times when Jim and I were the only two human beings at wolf camp. We'd spend our days out in the deep snow and biting cold, Jim filming and me recording sound, following the pack and watching their lives unfold. The sun would dip behind Williams Peak just after two in the afternoon. When the light grew too poor for filming, we'd go into the yurt, shed our heavy coats and boots, light the woodstove, put a homemade pizza in to cook and open a bottle of wine that we dubbed "Chateau Yurt." Then Jim and I would sit, talk about the behavior we had witnessed that day, and brainstorm scenes that we wanted to film in the days to come. At about 9 p.m. we'd head down the steps from the yurt platform to our sleeping tent.

Motaki, the first omega

Jim would light hurricane lamps and a small woodstove, just enough to take the edge off the cold, and we'd snuggle into bed under a heap of down.

Everything in our tent froze solid the moment the little woodstove went out. Sometimes I'd wake in the middle of the night, but I wouldn't dare move lest I disrupt the warm cocoon of our bed. So I would just lie still and listen to the forest at night. In the ice-cold air every sound seemed amplified, and I could hear the wolves stirring quietly just outside.

The pack kept its own schedule, but they often liked to bed down for a snooze at about the same time we did. We noticed that each wolf had its favorite sleeping spot. Just as dogs do, they turned a few circles before lying down. Repeating this action, night after night, each wolf would pack down a little hollow in the snow that we called a "sleep sink." After a while they would customize their sinks with willow branches and even pieces of elk hide, which may have added a layer of insulation. Once in bed, they never slept for more than a couple of hours at a stretch.

Sometimes, before going to sleep, I'd set up my microphone outside the tent and run the cables inside to the tape deck, which I'd have to keep in bed with us to prevent the batteries from dying in the cold. If the wolves awakened me during the night, all I had to do was push the record button, then I'd lie comfortably in bed and listen to their sounds through my headphones. I could hear the crunch of snow under their paws as they moved about, chewing on a leftover bone, conversing and playing just as they did during the day. They were completely unfazed by the darkness, and they didn't seem to mind the brutal cold one bit. On special occasions of their choosing, their howls would pierce the night.

Wahots slept as close to our tent as he possibly could. During the day he was somewhat aloof, but at night he liked to be near us. Perhaps the glow of our lamps and the muffled sound of our conversation were a comfort to him, reminding him of the days when we cared for him as a newborn pup. With only chain-link and thin canvas between him and us, we could hear his breathing just a few feet away. When he was inspired to howl in the middle of the night, he could launch us right out of bed.

Kamots slept farthest from us, and I believe that was by design. As alpha, he was the pack's guardian and protector, always ready to face danger. He seemed to understand that our camp was a safe place, so he stationed himself at a far outpost where he could keep eyes and ears on the rest of his territory. When he would howl, the sound would float in with a spooky echo coming off the mountains.

Wolves howl for more reasons than we will ever know, but after years in their company I began to detect nuances in their delivery and understand their basic meanings. Sometimes only Kamots and one or two others would howl into the night, seemingly in response to a far-off sound too faint for me to hear—perhaps a distant coyote or the rustle of an elk in the forest. In this way, I could feel the tension hanging in the silence as they listened for a reaction. Sometimes I thought they might be checking to see if other wolves were out there.

At other times their night howls sounded more like their energetic daytime howls—the ones that signaled a pack rally. As unperturbed by the cold and darkness as they were, I could tell they were indeed gathering for a rally or a spirited round of play. I could hear the galloping of paws as they ran toward each other, followed by cheerful snarling, yipping, and Chewbacca sounds that always accompanied their good times.

By far my favorite of their nocturnal vocalizations was what I thought of as "check-in howls," because it really seemed like they were checking in with one another in the darkness. In what sounded like a friendly conversation, one wolf would howl, then pause, and another would reply, and sometimes a third would chime in. I imagined they were saying something like:

"Are you there?"

"I'm here. I'm fine."

"I'm fine over here too."

The final note would end and then echo back from the mountains and fade away. I would hear Wahots shift restlessly and let out a deep sigh as he lay back down. Then everything would be silent again.

Daytime always centered on work. Our first priority was capitalizing on the best light of morning and evening for filming. We filled the rest of the daylight hours with repairing equipment, chopping wood, shoveling the all-important path to the outhouse, or doing any of the hundred other chores that had to be done. But at night all that fell away. It was a time just to be still. At night I could reflect on the incredible reality in which I found myself—snuggled and warm, next to my partner in creativity and in life, in the most stunning wilderness in North America, surrounded by a pack of wolves that knew me and trusted me.

That trust and that intimacy with another being were the greatest gifts that Lakota, Kamots, Chemukh, and their family gave to Jim and me. In the years we lived with the Sawtooth Pack, we experienced flashes of discovery and occasional moments of great drama, but these are not what I remember most. What resonates far stronger in my memory is the privilege of being among them and sharing day-to-day moments of simple beauty: the gestures of compassion, the acts of humor and curiosity, and the displays of unbreakable pack bonds. I came to know every wolf in that pack as an individual with a personality as unique as any human's, yet each individual expressed devotion to the others with an openness and a sincerity that are, I think, beyond most humans.

RECENTLY I WAS RUNNING ERRANDS near our home and had NPR on, as I usually do, half paying attention to the stories and half wrapped up in my own thoughts, when an interview hooked my interest. Terry Gross was talking to Kenneth Lonergan, director of the film *Manchester by the Sea*, on the subject of grief. Lonergan was explaining something he had read about overcoming loss, and he made the distinction between two ways of facing the future: "moving on" and "moving forward." To move on is to push the past away, forget grief, ignore the lessons, and put everything behind you. To move forward is to hold on to some of the sorrow and pain, but in doing so to honor the memory of what you've lost and the wisdom that came from the experience. That's a harder, more rewarding path to take. I guess I would say moving on is cutting off the past while moving forward is allowing it to change you.

Though it has been years since our project ended, when I heard this interview my mind immediately went back to wolves. I've heard some people speculate that wolves live in the "eternal now," that they are unburdened by past and future, forever existing in the purity of the moment. Having lived with them, I'm not convinced. Of course I'm certain that wolves are far more grounded in the here and now than we are. It's a struggle for me to stay in the present, when every sight and sound sparks a million mental connections firing in every direction. I suspect wolves are not plagued by the endless mental chatter that clouds our human minds, but I think it's an oversimplification to think that wolves don't remember their past or ponder their possible future.

We've learned from watching them that wolves are curious and inquisitive and prone to exploration. What would have impelled a wolf to make a journey of hundreds of miles if not his own imagination conjuring images of new places and new experiences? From our friends, we've heard heart-warming reports of a young male wolf exuberantly dragging a bison skull over a mile to deliver it to his younger siblings. What could have driven this seemingly pointless endeavor other than his own expectation of the joyful reception his gift would elicit? We have also watched the Sawtooth Pack stand in silent mourning in the spot where, months earlier, their beloved packmate was killed. What could have been running through their minds other than memories of their lost friend and the grief of knowing she was gone?

I find myself believing that wolves do not live in the eternal now, nor do they blindly move on and forget past experiences. Wolves move forward. In fact, they may do so better than many of us can. Think of all we have done to wolves over the centuries, and all the suffering we have inflicted upon them. We have shot them and caught them in painful traps by the thousands. We have killed parents in front of their children and ripped families apart. As a result, wolves have changed. Where once they strode con-fidently over open plains, they now live in the shadows. They have adapted to living as fugitives in a world domi-nated by unfriendly human beings.

I believe wolves hold on to memories of grief and fear and carry them as they move forward. Perhaps they carry older, more pleasant memories too—memories of past times of camaraderie when wolves and humans were equal on this Earth. Even today, if you see a wolf in the wild, you can rest assured he knows you're there and he's letting you see him. Wolves are still curious, still playful and compassionate. Occasionally they still reach out to us. After all we've done, we have not broken them. In the long, tangled relationship between our two species, perhaps we are the ones that have failed to move forward, opting instead to simply move on. We have forgotten our past kinship and have slammed the door on these ancient memories.

In the closing paragraphs of one of the greatest books ever written on wolves, *Never Cry Wolf,* Farley Mowat writes: "We have doomed the wolf not for what it is, but for what we deliberately and mistakenly perceive it to be—the mythologized epitome of a savage ruthless killer—which is, in reality, no more than a reflected image of ourself."

What strikes me most deeply in this passage is the word "deliberately," for without a doubt our vilification of the wolf has been deliberate and methodical. Our ability to kill wolves with the dispassion that we do requires willful ignorance, given the intelligent, family-oriented creatures they have revealed themselves to be. Over centuries we have wrapped them in a complex mythology woven of medieval fears and superstitions. Wolves may indeed be killers, but then again so are we, and we are much more ruthless, efficient, and prolific at it.

If wolves were half as terrible as we make them out to be, they wouldn't have been so easy to eradicate. A primer on wolf trapping from 1909 includes a quote from a hunter that attests, "I never hesitate in entering a wolf den, even when I know the mother wolf is with her young, and have never known one to act vicious, but always sneaking and cowardly." The book goes on to instruct readers on how to use "strong fishhooks" to drag pups out of their dens. In the face of such brutality, adult wolves responded only with passivity and fear, and perhaps despair at being unable to protect their pups from such a powerful foe. Yet we continue to call them savage. Farley Mowat was right; some of what we hate about wolves is nothing more than our own reflection.

Wolves have never stopped showing us who they really are. Beneath the suspicion and mistrust that we heap on them, they really do offer a reflection of ourselves, as Mowat wrote, and it is far brighter than the one we've grown accustomed to seeing. To my eyes, wolves are virtual superheroes, embodying so many of the qualities we admire in the best of our kind. As athletes, they are unparalleled, able to pull down animals more than four times their size with nothing but their legs and their jaws. As team players they excel at cooperation, reading each other's actions and thinking as one. They look after their youngest and oldest with limitless patience, care, and devotion. Almost every day in a wolf's life requires some kind of heroism, some act worthy of our admiration. I suppose when we lived more like they lived, we were heroic too. Perhaps we resent them for retaining some of the virtues we've lost.

Our last film, *Living with Wolves*, aired in 2005, just as films like ours were falling out of favor with the major networks, replaced by so-called reality programs that were cheaper in every sense of the word. The reality-style competition shows that emerged during that era spawned a phrase uttered with such frequency that it became a cliché: "I'm not here to make friends." In the world of reality television, compassion is a liability. Might is passed off as strength, and selfishness masquerades as clear-eyed realism. Meanwhile, we seem to mistake gentleness for weakness. It has been said that a society can be judged by how it treats its most vulnerable. Maybe we're conflicted about wolves because we're so conflicted about ourselves.

IN THE END, THE ONLY VIRTUE that wolves need from us is honesty—regarding them, regarding us, and regarding our shared past. Only by seeing them as they are, as neither demon nor deity but as creatures worthy of our admiration, will we find tolerance within our own human character. Wolves still have enough land to claim as home, enough prey to hunt, enough space to explore, but we have to let them. Ed Bangs is a retired U.S. Fish and Wildlife official who led the wolf reintroduction effort in the northern Rockies. For 23 years he was at the center of the storm, caught in the middle between wolf haters and wolf lovers. He put this ultimate truth as simply as anyone could when he wrote, "I've always said that the best wolf habitat resides in the human heart. You have to leave a little space for them to live."

Kamots, Lakota, Matsi, and the other wolves of the Sawtooth Pack have done their part and told their stories to the world. Wolves across America have gradually revealed the depth and complexity of their lives to any willing to see. Nothing more should be expected of them. It is up to us now.

ACKNOWLEDGMENTS

I T'S BEEN ALMOST 30 YEARS since our project with the Sawtooth wolves began. This book could not have been written during the years when the wolves were part of our daily lives, or even in the years immediately following. Sometimes the seeds of enlightenment need time to germinate.

We are grateful for the friendship and talent of James Manfull, the creative writer of our film treatments and books about the Sawtooth Pack. He has worked tirelessly, helping us shape this deeply personal story and share our journey with others.

We are thankful to our editor, Susan Hitchcock, who has always believed in our work and our story and who encouraged us to write this book about the gifts of wisdom the Sawtooth Pack taught us. With her at National Geographic Books, we thank photo editor Susan Blair, designer Nicole Miller, and production editor Judith Klein for helping make this book a reality.

We owe special thanks to Garrick Dutcher, the multitalented and creative program director of our nonprofit organization, Living with Wolves. His hard work and dedication are equaled only by his passion for wolves.

We are indebted to Marc Bekoff for his generosity in agreeing to write the foreword to this book. Through his research and many books, Marc has singlehandedly brought

Amani on a crisp autumn morning

the emotional lives of animals to the forefront. No longer can people talk about animals without acknowledging the emotional lives they lead.

We would also like to thank our dear friend Norma Douglas, who has always been with us throughout our journey with wolves. Whether on our films, our books, or our nonprofit organization, Norma has always "had our back" and has been instrumental in helping us in whatever we set out to do.

In this book, in addition to the stories about our time with the Sawtooth Pack, it was important for us to share stories of many wolves we don't know personally: the wolves of Denali National Park and Preserve and in Yellowstone National Park. These stories would not be as rich if it were not for the help and dedication of Rick McIntyre, who has kept a watchful eye on the Yellowstone wolves since they were reintroduced to the park, and of his many wolf watchers, too numerous to name, who are out there in the field day after day, rain or shine, extreme heat or bitter cold. We recommend that if you are ever in Yellowstone, you stop by to say hello to these volunteers. You can always find them parked on the side of the road with their spotting scopes!

We would like to thank Wally Montgomery for the many Yellowstone stories and the video footage he has shared with us, and Nell and Bob Harvey for taking the time to share their stories of watching wolves in Yellowstone as we stood together on top of a rise on a very cold May morning watching the Junction Butte Pack and their pups.

ACKNOWLEDGMENTS

Thank you to Doug Smith for sharing so much of his knowledge and for always taking the time to speak with us, and to Kira Cassidy for her groundbreaking research on the importance of older wolves within a pack. It's a wonderful thing to meet someone new whose theories, feelings, and ideas resonate so well with our own. And of course, we recall with gratitude the memory of Gordon Haber, who was able to express so well the emotional lives of the wolves he was watching in Denali for so many years. We feel very lucky to have spent the time we had with him in the field. He was ahead of his time, and his life was cut too short.

Most of all, none of this would have been possible without the wolves of the Sawtooth Pack. They gave us the greatest gift that could ever be given from one species to another— trust—and with that trust they opened their world to us. In the end, they made us better for it. We became better human beings by watching and learning what the wolves had to teach us about life, love, compassion, and care. We will never forget them or what they taught us. They are with us every day, in everything we do, guiding us forward.

ABOUT THE AUTHORS

JIM AND JAMIE DUTCHER ARE universally recognized as two of America's most knowledgeable experts on wolves. With nearly 30 years of their lives devoted to eradicating ancient myths surrounding wolves, they are dedicated to sharing their unique and personal eyewitness knowledge.

During the 1990s, they lived in a tented camp in the Sawtooth Mountains of Idaho, where they intimately observed the social hierarchy of the now famous Sawtooth Pack. Their six-year experience led to the creation of the National Geographic Society book *The Hidden Life of Wolves,* as well as five other books and three prime-time television documentaries on wolves.

In 1962 Jim Dutcher began his career as an underwater cinematographer and then a wildlife film producer for National Geographic. Combining extraordinary camera work and the trust he gains from his subjects, Jim's films have taken audiences into a world never before filmed: inside beaver lodges, down burrows to observe wolf pups, and into the life of a mother mountain lion as she cares for her newborn kittens. His work includes the National Geographic special *Rocky Mountain Beaver Pond* and ABC

Jim and Jamie Dutcher

World of Discovery's two highest-rated films, *Cougar: Ghost of the Rockies* and *Wolf: Return of a Legend.*

In 1991 Jim received the prestigious Wrangler Award for his cougar documentary from Oklahoma's National Cowboy Hall of Fame. In 1995 the governor of Idaho appointed Jim as an ex officio member of the Idaho Wolf Management Committee, where he served until 2001. As part of the wolf reintroduction initiative at Yellowstone National Park, Jim served as a consultant for the design of the pack holding enclosures and as a specialist in the handling of the reintroduced wolves brought from Canada.

Jamie Dutcher began her career as an animal keeper and veterinary technician at the Smithsonian's National Zoo in Washington, D.C. She contributed her knowledge of animal husbandry and medical care to Jim Dutcher's film projects and recorded the vocalizations of the Sawtooth Pack, winning an Emmy. Working together, Jim and Jamie created two of Discovery Channel's highest-rated wildlife films, *Wolves at Our Door* and *Living with Wolves*. Their television specials on wolves have won three Primetime Emmys, for cinematography, for outstanding programming, and for sound recording. The Dutchers also led three National Geographic expeditions to Alaska, working with revered wolf biologist Dr. Gordon Haber, observing pack hunting techniques and the culture of shared knowledge within individual wolf families.

In 2006, moved by the plight of wolves, the Dutchers put down their filmmaking equipment and founded the nonprofit organization Living with Wolves, which is dedicated to raising public awareness of the truth about wolves, their social nature, their importance to healthy ecosystems, and the threats to their survival. Since then, they have presented trusted, factual information through social media, books, educator guides for schools, films, and museum exhibits to more than 30 million people worldwide, becoming the premier advocates for wolves, bringing understanding of this misunderstood and persecuted species by sharing their knowledge and lifetime of study. They have appeared at the New York's American Museum of Natural History, Harvard University, Chicago's Field Museum, the California Academy of Sciences, the Smithsonian Institution, and the Museum of Natural History of London. They have been interviewed for radio, television, and print media including the *New York Times,* the *Washington Post,* PBS, NPR, BBC, *Today,* and *Good Morning America.* Their personal multimedia wolf presentation is a featured part of the National Geographic Live! programming.

MARC BEKOFF is professor emeritus of ecology and evolutionary biology at the University of Colorado Boulder. He has won many awards including a Guggenheim Fellowship for his research in animal behavior and behavioral ecology. His latest books include *Rewilding Our Hearts: Building Pathways of Compassion and Coexistence* and *The Animals' Agenda: Freedom, Compassion, and Coexistence in the Human Age.*

ABOUT
Living with Wolves

WITH UNIQUE EXPERIENCES and personal observations of wolves driving their concern, Jim and Jamie Dutcher continue to devote their lives to changing deadly misunderstandings about these keystone animals. Putting aside their award-winning film career, they founded the nonprofit organization Living with Wolves in 2006.

Dedicated to raising broad public awareness of the truth about wolves, their social nature, their importance in healthy ecosystems, and the threats to their survival, Living with Wolves fills a unique position as the leading national advocacy organization devoted solely to wolves. It provides a wide range of trustworthy and effective actions, building tolerance that can lead to coexistence between people and wolves sharing land.

Honorary board members Jane Goodall, Barry Lopez, and Robert Redford join an esteemed board of directors, all eminently qualified to oversee this wide range of outreach activities. Importantly, an impressive advisory board provides expert guidance in fields that include biological sciences, wolf recovery, ranching and livestock, the

Jim with Matsi and Kamots

economics of wolf reintroduction and tourism, and ethical hunters who see the value in sharing the land with predators. Also important, the Dutchers' long-standing collaboration with National Geographic continues to enable them to bring a wide range of programming to a constantly growing and diverse national audience.

The ongoing partnership with National Geographic has produced works in print that include *The Hidden Life of Wolves,* a book now available in six languages, and two award-winning books for children, *A Friend for Lakota: The Incredible True Story of a Wolf Who Braved Bullying* and *Living With Wolves!*

Also produced in collaboration with National Geographic, online educational guides for families and educators provide teachers, from kindergarten through high school, with activities, maps, and multimedia resources to enrich students' understanding of wolves and their role in the natural world.

Jim and Jamie continue personally to present multimedia events about wolves in conjunction with National Geographic Live! to auditoriums of adults and children eager to learn.

Garrick Dutcher, program director for Living with Wolves, provides valued in-person reports to state, regional, and federal officials who decide the fate of wolves and who request trustworthy information as legislation is under consideration. Key to this important aspect of Garrick's work is our funding of field research in Yellowstone National Park, Grand Teton National Park, and Denali National Park and Preserve.

Living with Wolves' website and social media on Facebook and Twitter provide a source for accurate and factual information, read by many thousands of people who strive to bring a commonsense, fact-based, workable, and ecologically healthy approach to the forefront of the wolf controversy.

An extensive photography exhibition created by Living with Wolves to bring the world of the wolf to more audiences has been displayed at the Russell Senate Office Building in Washington, D.C., at Chicago's Field Museum, and at the Detroit Zoo.

To learn more about Living with Wolves, we invite you to visit us online at *www.livingwithwolves.org.*

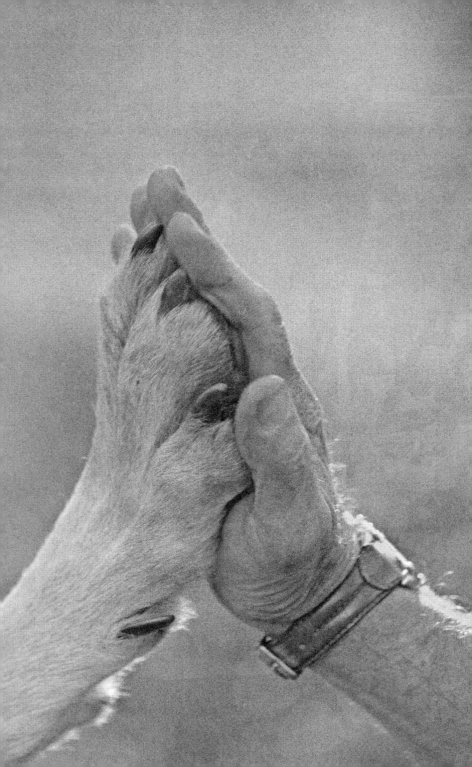

AN INCREDIBLE TRUE STORY

You've read their story—
now discover intimate
photographs of the
Sawtooth wolf pack by
Jim and Jamie Dutcher.

NATIONAL GEOGRAPHIC

THE HIDDEN LIFE OF
WOLVES

JIM AND JAMIE DUTCHER   FOREWORD BY ROBERT REDFORD

AVAILABLE WHEREVER BOOKS ARE SOLD
AND AT NATIONALGEOGRAPHIC.COM/BOOKS

NATIONAL
GEOGRAPHIC

NATGEOBOOKS   @NATGEOBOOKS   © 2018 National Geographic Partners, LLC